Living by the Rules of the Sea

Living by the Rules of the Sea

David M. Bush, Orrin H. Pilkey Jr., and William J. Neal

Duke University Press   Durham and London 1996

To Carmen, Sharlene, and Mary

© 1996 Duke University Press
All rights reserved
Printed in the United States of America
on acid-free paper ∞
Typeset in Sabon.
Library of Congress Cataloging-in-
Publication Data appear on the last
printed page of this book.

Contents

List of Figures and Tables

Figures

Tables

Preface

We do not recommend living on barrier islands and we definitely would not want our loved ones to live there. The hazards are numerous and difficult to avoid by evacuation. Development on these islands is destroying a critical and limited ecosystem. Read on, however, if you are a gambler and you'd like to take your chances. This book will help you reduce your risk. We want to convince you to live by the rules of the sea.

When a storm strikes a barrier island, the typical reaction is to clean up and rebuild. All kinds of government resources pour into the area to help the victims of the natural disaster. Even patriotism swells. After Hurricane Hugo crossed the South Carolina shore, American flags appeared everywhere, sort of a symbol of the courage and determination of the local people to bounce back.

South Carolina did bounce back, but as it turns out, Hurricane Hugo may have been a giant urban renewal project. Destroyed homes were replaced by bigger ones, now without dune protection and a bit closer to the surf zone than their pre-Hugo predecessors. As a result of Hurricane Hugo more property than ever now stands at high risk of destruction.

There was another way. South Carolinians could have paused to reflect on the wisdom of building near the beach. If they had looked a little closer they might have observed that some buildings survived almost unscathed while some of the neighboring structures disappeared completely. Pursuing it further they could have recognized that the surviving buildings were often located at higher elevations and/or were within dense protective forests. They could have observed and learned the simple lesson from Mother Nature: do not rebuild destroyed structures in place.

Unfortunately, South Carolina's islands (as well as those in every other coastal state) were developed no differently than the wheat fields in Kansas or corn fields in Illinois. Not the slightest attention was paid to the processes that were bound to occur in the next storm. South Carolina did attempt to recognize in its laws the threat of the eroding shoreline to beachfront buildings. But that precaution was substantially beaten back by the Supreme Court's famous *Lucas* decision, which required the state to purchase property where it had prohibited construction.

There is another way, involving better recognition of the events that are unique to barrier islands and coastal living. The Kansas wheat field development mentality, where market forces rather than nature's forces determine development patterns, should have come under question. Survivors could have reflected on the impact of historical storms and post-storm redevelopment and avoided repeating the same mistakes. Islands could even have been "repaired" to enhance the protective aspects of natural island environments, protective aspects too often destroyed or diminished in the name of progress and development.

We have been privileged to experience up close coastal processes such as storms and shoreline retreat. For Pilkey it began with the destruction of his parents' home in Waveland, Mississippi, by Hurricane Camille (1969). In more recent years we have taken a first-hand look at Hurricanes Gilbert, Hugo (in the Virgin Islands, Puerto Rico, and South Carolina), Andrew, and Emily and their impact on development. Over the years we have mapped and studied most of the barrier islands in the United States in the process of writing or editing the Living with the Shore series published by Duke University Press. In these books, now numbering 19, we examine, in site-specific fashion, coastal hazards of all kinds, ranging from the tsunami threat to Homer Spit, Alaska, to the subsidence-caused shoreline retreat problem of the Louisiana barrier islands, to the relentless 6-foot-per-year shoreline retreat rate on South Nags Head, North Carolina, where in some locations building row 3 is now beachfront property.

As we look around the country, especially at barrier island development, we find that, most commonly, property damage mitigation is carried out by zoning (land-use planning) or by engineering (e.g., construction of seawalls). In this volume we present a simple and new view of the problem of property damage mitigation, from a geologic and oceanographic viewpoint. Our approach is a simple, four-step process:

(1) understand the physical processes (these are the "hazards"); (2) map zones of relative risk of property damage caused by these processes; (3) develop site-specific (and nonstructural) property damage mitigation techniques; and (4) implement mitigation techniques. We understand that land-use planning is the main way to implement our recommendations and that this is a highly politicized process. Even in this extremely hazardous environment, the rush to develop will tend to make implementation of new ideas very difficult. Nonetheless we have to start somewhere, and there is no better time than the present.

We hope that this volume will be useful to individual property owners and to those seeking to purchase barrier island property. We hope we provide the basis for making the wisest possible decision in terms of minimizing the potential for future property damage.

We hasten to repeat, however, that we do not recommend or endorse the idea of living on a barrier island. Applying the stiffest of standards (Where would you recommend that your aging retired parents purchase property?) we strongly recommend against barrier island property purchase. Better to choose a high-elevation inland site. None of us owns property on a barrier island, and we hope our loved ones never do.

This book is a product of long-term support of the Living with the Shore series by the Federal Emergency Management Agency. We are extremely grateful for this help. Initial funding for the Property Damage Mitigation project was provided by the Federal Emergency Management Agency (FEMA) through grant EMW-G-0010 to Duke University and by the National Oceanic and Atmospheric Administration (NOAA) through grants NA89AA-D-CZ064 and NA89AA-D-CZ218 to Duke University. Continued funding has been provided by FEMA through grant EMW-G-90-G-3498 to Duke University.

So many people played a role in getting this study off the ground and moving it forward that it is almost impossible to list them all. Many thanks in the beginning to the NOAA personnel who helped us design and nurture our original ideas for the project; to Jim Burgess and, especially, Dave Kaiser for answering a thousand or more questions. The FEMA staff in Washington and Atlanta were in on the original property damage mitigation ideas. Jane Bullock, Gary Johnson, Fred Sharrocks, Dick Krimm, Gene Zeizel, Art Zeizel, Glen Woodard, Bill Massey, Todd Davison, John Gambel, John McShane, and others we have probably neglected (to whom we apologize) have walked us through much of this. All field trip participants gave something of themselves to this project, providing constructive comments for discussion and planting the seeds for further work.

The Coastal Zone managers and personnel of the participating states deserve special mention for allowing us to further complicate their lives. Many thanks to Directors Dave Owens, George Everette, and Roger Schecter of the North Carolina Division of Coastal Management; Executive Director H. Wayne Beam of the South Carolina Department of Health and Environmental Control, Office of Ocean and Coastal Resource Management (OCRM, formerly the South Carolina Coastal Council); Jim Stoutamire of the Florida Department of Environmental Regulation; and Mike McDonald and Frank Koutnik of the Florida Department of Community Affairs, Division of Emergency Management.

The conclusions of this book, however, are those of the authors, based on various published reports and studies of record, and are not meant to reflect the views of any specific agency. Support for the Risk Mapping Project was also provided by the Insurance Institute for Property Loss Reduction and State Farm Insurance.

The mapping and much of the description in chapter 9 were done by Matthew Stutz (Southampton, New York, and Galveston, Texas) and Amy Reesman (Fernandina and Venice, Florida). Gered Lennon also helped with field work in Galveston. Special thanks for all their hard work and superior input.

Amber Taylor drafted, redrafted, or oversaw drafting of the risk maps and line drawings. Thanks to Debbie Gooch for typing several drafts. Kathie Dixon, Craig Webb, Rob

Thieler, Rob Young, Rodney Priddy, Jay Williams, Kay Carlson, Jackie Howard, Mitchell Malone, Andy Coburn, Susan Bates, Art Trembanis, Carmen Bush, Sharlene Pilkey, Mary Neal, Tsung-Yi Lin, and Yarta Clemens Major helped with many tasks along the way, including assistance with a great deal of the primary research and writing. Andy Coburn made major contributions to the chapter on hazard mapping, and most of the fieldwork and preliminary mapping was his. Each of these capable people was employed for a time by the Program for the Study of Developed Shorelines and all contributed their own special insights and enthusiasm to the project. Richard Lefebvre helped develop the coastal risk assessment concept diagram. The Pandora's Island drawings were done by Charles Pilkey.

Finally, this project would be nowhere without the kind and exceptional help of three very special people. Jane Bullock of FEMA, Rich Shaw of the North Carolina Division of Coastal Management, and Gered Lennon, then of the former South Carolina Coastal Council, gave us professional guidance, moral support, and friendship both in the field and in the office. They made this a fun as well as a worthwhile project.

David M. Bush
Orrin H. Pilkey Jr.
William J. Neal

Living by the Rules of the Sea

1 Bigger Disasters Building to Happen

In 1954 Hurricane Hazel raked the coast of the Carolinas, a storm not unlike the hundreds of hurricanes through the centuries, but she left a path of over $280 million in destruction (fig. 1.1). Hurricane Betsy (1965) became the first hurricane to exceed $1 billion in total damages, mostly in Florida and Louisiana (fig. 1.2); in 1969 Hurricane Camille killed 256 along the Gulf coast and slightly exceeded Betsy in damages, accounting for over $1.4 billion in property loss. Ten years later Hurricane Frederic ran up a $2.3 billion bill in losses along much of the same Mississippi-Alabama coast (fig. 1.3). That record stood until 1989 when nature's forces turned again on the Carolinas and Hurricane Hugo left a wake of widespread destruction along the coast as well as inland to the tune of $7 billion. And the cost more than tripled again in 1992 when Hurricane Andrew's visit to Florida and Louisiana destroyed or damaged property worth about $25 billion (table 1.1). The ten most costly hurricanes of the twentieth century came after 1964. Is each successive hurricane getting bigger, more severe in wind velocity, wave energy, and associated flooding?

Hurricanes are not the only big events in the spotlight. Winter northeasters expend their energy over great stretches of the Atlantic coast and wave erosion associated with extratropical storms is not uncommon on the Gulf shores as well. The Ash Wednesday storm of 1962 affected nearly the entire U.S. Atlantic

a   b   c

1.1 Hurricane Hazel (1954); Long Beach, NC (a) before the storm; (b) right after; (c) twenty years later—more houses and development, a bit closer to the shoreline than before, awaiting the next hurricane. Photos courtesy of the Wilmington District, U.S. Army Corps of Engineers.

coast, causing over $300 million in property damage, changing the shape of the shore in terms of beach and dune erosion, opening new inlets, and burying roads and property under overwash sand (fig. 1.4). Extensive overwash was again the case in North Carolina during the 1972 Lincoln's Birthday storm. The Halloween storm of 1991 was described as one of the greatest northeasters in 50 years, but just two years later, a 1993 northeaster was dubbed the "Storm of the Century"! Are big winter storms also becoming more frequent?

No matter how we measure the impact, whether in number of buildings destroyed, lives lost, people temporarily displaced, or the billions of dollars in losses, storms are becoming an ever-greater drain on our resources. We have to wonder what is happening and who is paying the bill for all of these losses.

Although there are suggestions that storm frequency and intensity do vary with global climatic changes, today's huge losses are not being caused by bigger hurricanes or winter storms. Hurricanes Andrew and Hugo were not as powerful as Hurricane Camille, yet their financial impacts were far greater. Hurricane Agnes (1972) was a lightweight but is fourth on the all-time damage list (see table 1.1). On the other hand, the work of Davis and Dolan (1993) suggests that in the past decade there have been fewer northeasters but

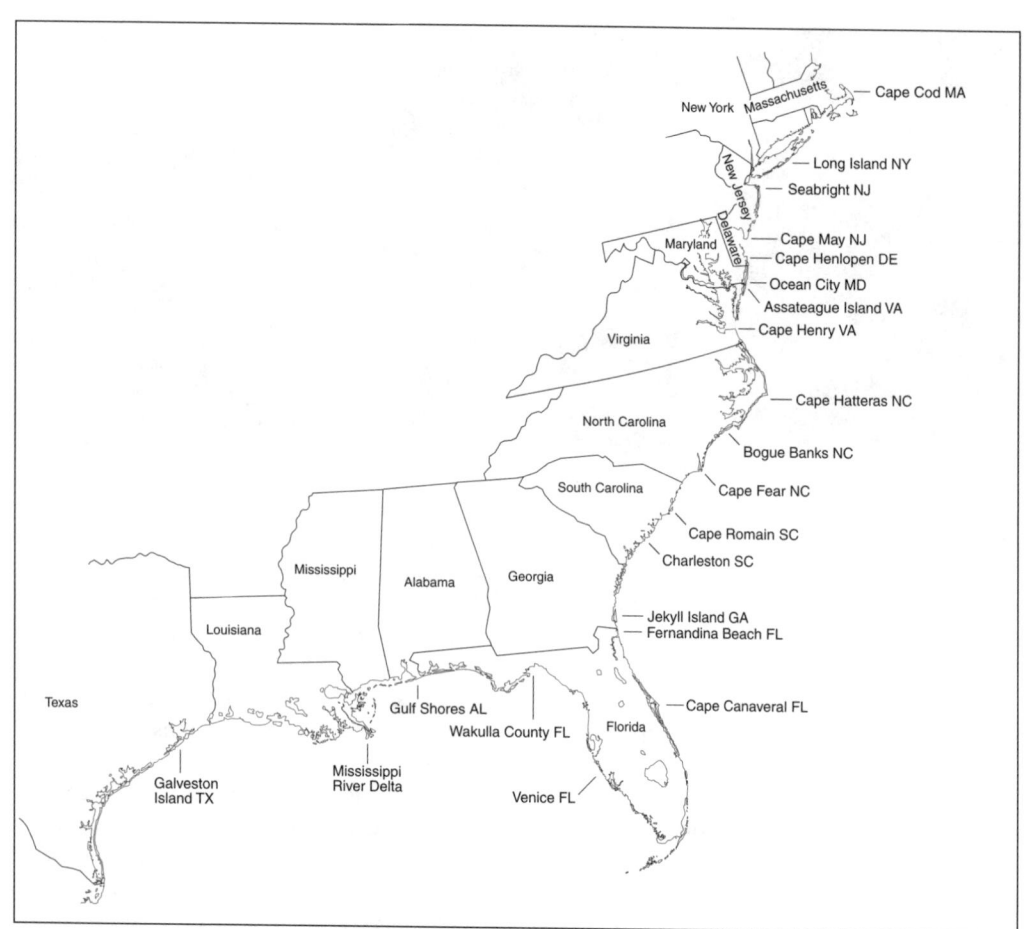

1.2 Index map of the U.S. Atlantic Ocean and Gulf of Mexico coasts showing some of the coastal communities discussed in text.

1.3 Aerial photo of Dauphin Island, AL, taken immediately after Hurricane Frederic (1979). Note the "overwash fabric" on the island produced as storm-surge waters first washed lagoonward across the island and then, as wind direction reversed, came back and washed across the island in a seaward direction. The location of the human-built canals focused erosion, leading to their breeching. Photo provided by the Topographic Bureau of the Florida Department of Transportation.

more big ones than in the prior 50 years of record. If the latter is a long-term trend, what are the implications for development in the coastal zone?

What has changed most significantly since the 1950s is the dramatic, rapid population increase in the coastal zone. The associated density of development is in an area that is far more vulnerable than inland areas to being impacted by natural processes such as wind, waves, storm-surge flooding, coastal erosion, and related processes.

Who is paying the ever-increasing costs of these events and of the day-to-day losses from "little" events that go unreported? We are all paying, and in a variety of ways. The individual property owners suffer the obvious loss; their neighbors and community maintain the streets and services that must be repaired and replaced; the county and state taxpayers share the burden of such costs; every federal taxpayer (even those far removed from the shore) also contributes through underwriting disaster assistance, the national flood insurance program, loans to communities to build and replace the infrastructure of services, loans to businesses, loans to veterans, temporary housing for those who are displaced—and the list of "donuts for disaster victims" goes on. Taxpayers at all levels support the federal portion of costs for U.S. Army Corps of Engineers projects to rebuild beaches and to construct coastal defenses such as seawalls that protect property at the cost of beach loss (a public resource). When we pay "tourist" taxes—the

Table 1.1 Costliest Hurricanes in U.S. History, in Order of Unadjusted Cost

Rank	Hurricane	Year	C	Damage in millions of $	Rank	Hurricane	Year	C	Damage in millions of $
1	Andrew (FL, LA)	1992	4	25,000	22	David (FL, e U.S.)	1979	2	320
2	Hugo (SC)	1989	4	7,000	23	Iwa (Kauai, HI)	1982	2?	312
3	Frederic (AL, MS)	1979	3	2,300	24	New England*	1938	3	306
4	Agnes (ne U.S.)	1972	1	2,100	25	Kate (FL Keys, nw FL)	1985	2	300
5	Alicia (n TX)	1983	3	2,000	26	Allen (s TX)	1980	3	300
6	Iniki (Kauai, HI)	1992	3?	1,800	27	Norman (CA)	1978	t	300
7	Bob (NC, ne U.S.)	1991	2	1,500	28	Hazel (SC, NC)*	1954	4	281
8	Juan (LA)	1985	1	1,500	29	Dora (ne FL)	1964	2	250
9	Camille (MS, AL)	1969	5	1,421	30	Beulah (s TX)	1967	3	200
10	Betsy (se FL, se LA)	1965	3	1,420	31	Kathleen (CA, AZ)	1976	T	160
11	Elena (MS, AL, nw FL)	1985	3	1,250	32	Audrey (LA, n TX)	1957	4	150
12	Gloria (e U.S.)*	1985	3	900	33	Carmen (LA)	1974	3	150
13	Diane (ne U.S.)	1955	1	832	34	Cleo (se FL)	1964	2	128
14	Allison (n TX)	1989	T	500	35	Hilda (central LA)	1964	3	125
15	Eloise (nw FL)	1975	T	490	36	FL (Miami)	1926	4	112
16	Carol (ne U.S.)*	1954	3	461	37	Unnamed (se FL, LA, MS)	1947	4	110
17	Celia (s TX)	1970	3	453	38	Unnamed (ne U.S.)*	1944	3	100+
18	Carla (TX)	1961	5	408					
19	Claudette (n TX)	1979	T	400					
20	Donna (FL, e U.S.)	1960	4	387					
21	Olivia (CA)	1982	t	325					

*=Moving more than 30 mph.
C=Hurricane Category (Saffir/Simpson scale, Table 3.2).
T=Only of tropical storm intensity but included because of high cost of damage.
t=Only of tropical depression intensity but included because of high cost of damage.
Source: Adapted from Hebert and Taylor, 1988.

insurance company, you will share the cost of coastal property losses. Insurance companies with home offices outside of the storm area, for example, in Connecticut or Michigan, will take the financial hit from the resulting claims from damaged buildings in Texas or wind-ripped mobile homes in Florida. Hurricane Andrew caused the demise of nine Florida-based insurance companies and caused several more to withdraw or reduce their exposure in Florida. Ownership in any business closed down or lost in a storm is ownership in the loss. (Surviving building suppliers and ice plants do a land-office business in the aftermath.)

All of us pay in one way or another for the studies, plans, and projects to restore, maintain, and protect these high-risk coastal communities. As residents, recreational users, taxpayers, investors, insurance customers, and so on, we are the underwriters of coastal development, coastal land use, and mitigation against coastal hazards. We are contributing to a giant welfare system in which high-risk development is encouraged and then rewarded when disaster strikes!

Coastal Population Bomb

Despite the all-too-real potential for coastal catastrophe, the U.S. population is migrating toward the coast at a rapid rate. In recent decades, areas within 5 miles (8 kilometers) of

1.4 Harvey Cedars, NJ, after the 1962 Ash Wednesday storm. A new inlet has formed—its location determined by the bulkhead on the lagoon side of the island. Clearly this inlet formed by the storm-surge ebb, focused by the bulkhead, rushing across the island from the lagoon to the sea. Photo courtesy of U.S. Army Corps of Engineers.

motel fee, the beach user's tag, the community sales tax—we contribute to the maintenance of the seawall or the next truckload of sand purchased for the town beach.

If you purchase or own property insurance, especially flood insurance, or own stock in an the shoreline have experienced population growth rates three times the national average. Unfortunately, the demand for ocean views and private beach access encourages development in extreme-risk beachfront areas, placing ever-increasing property investments and more and more residents at high risk to the impacts of storm winds, waves, and flooding. Oceanfront property is at risk from natural shoreline migration.

The storm record demonstrates that the danger is not just to beachfront property, but extends inland, especially on barrier islands. As more and more people live and vacation along the coast, more lives, property, and dollars are put at risk. More private development in the coastal zone means more limited access to beaches. The growing congestion, in turn, means that evacuation capabilities may be exceeded. Risks increase with growth. The end result is the spiraling total value of coastal property and the staggering exposure to potential losses in storms.

A survey and projections of U.S. coastal population change by the National Oceanic and Atmospheric Administration (Culliton et al., 1990) gives a good idea of the magnitude of the growth. In the United States there are 30 coastal states, including those along the Great Lakes. Counting the District of Columbia, 23 boroughs or census areas in Alaska, and 42 independent cities in Virginia and Maryland that are "county equivalents," there are 451 coastal

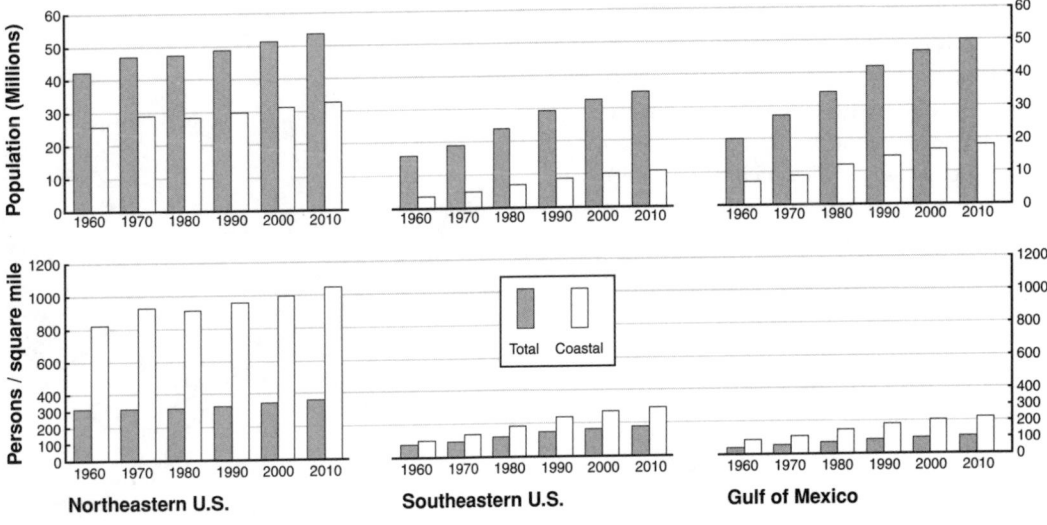

counties and 1,569 noncoastal counties in the 30 coastal states. In 1960, 80 million people lived within U.S. coastal counties (Culliton et al., 1990). By 1990 the figure had grown to 110 million. By 2010, projections are for over 127 million people to live in U.S. coastal counties (fig. 1.5)! These numbers amounted to a population concentration of 1,177 persons per shoreline mile (706 per shoreline kilometer) in 1988, with 1,358 per mile (814 per kilometer) projected by 2010. Increasing concentration of coastal population and vulnerable development is especially alarming because of a projected increase in hurricane frequency and intensity created by changing cyclical climatic

1.5 Growth of the U.S. coastal population in northeastern U.S., southeastern U.S., and Gulf of Mexico states. The vast increase in coastal population and coastal population density certainly will lead to ever-increasing property damage in storms.

patterns (Gray, 1992) and possibly the greenhouse effect. Even without such changes, a reflection on the history of hurricanes and northeasters for the Atlantic and Gulf coasts tells us that nearly 40 million people's lives are going to be affected one way or another in the next few decades. How do we defuse this population bomb?

Hazards, Vulnerability, Risk, and Mitigation

As long as coasts have existed, waves, tides, hurricanes, northeasters, tsunamis, and all of the associated natural changes (see chapter 2) have come and gone without notice. These natural processes, which include erosion, flooding, high winds, and similar high-energy events, only became hazards (or hazardous) when humans occupied the coastal zone with fixed habitations and related structures such as docks, wharfs, buildings, pipelines, utilities, and roads. A natural *hazard* is any natural physical process with the potential to cause loss of life or property (e.g., thunderstorms, river floods, earthquakes, volcanic eruptions, hurricanes). Natural processes have always been hazardous, but there haven't always been people and development *vulnerable* to the hazards. Without both hazards and vulnerability, there is no *risk*. A house on the floodplain is vulnerable, susceptible to flooding; the risk can be thought of as the potential or probability of damage or loss from the inevitable flood. Another way of looking at risk is to put it in terms of how much you are willing to lose if you locate in a hazard zone.

What's Wrong with This Island?

Poor development siting and inappropriate island alterations are schematically illustrated in figure 1.6. Also shown are examples of the different types of coastal hazards associated with

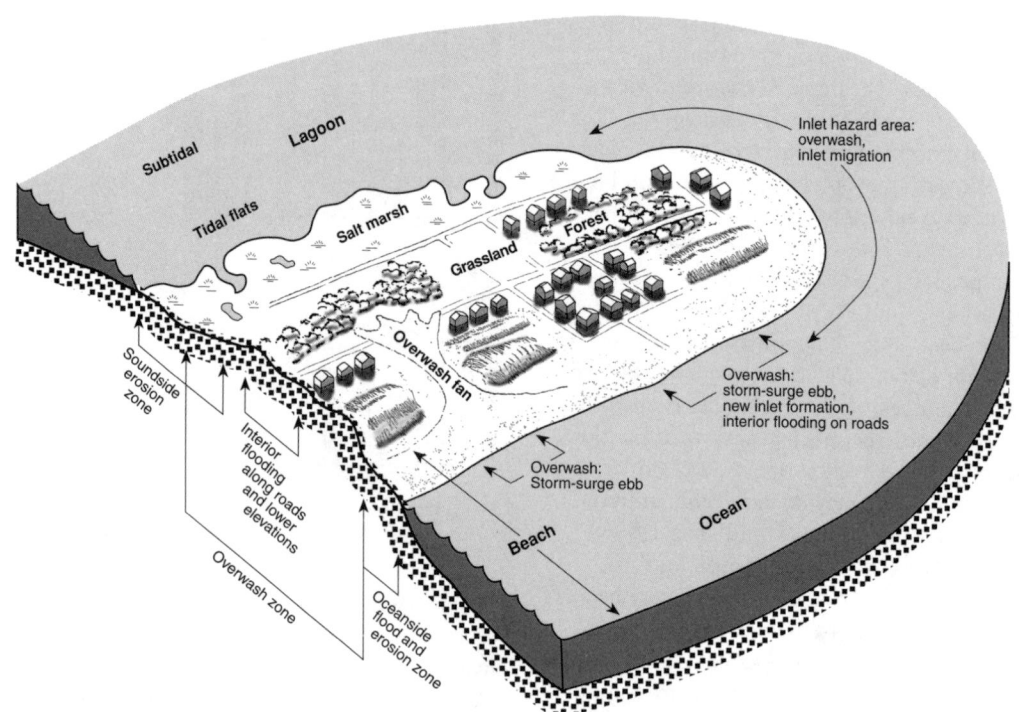

different areas of the island. In this imaginary development, single-family houses are unwisely located in island interior flooding zones, in an inlet hazard area, and directly on the ocean beach. A large area of dunes has been removed for siting of houses, increasing the potential for damage from overwash, storm-surge ebb, new inlet formation, and interior flooding. Maritime forest has been removed for other houses. Houses located on high

1.6 Cartoon showing barrier island environments and associated hazards, plus examples of typical barrier island development. Note areas where frontal dunes, interior dunes, and forest have been cleared for siting of houses and roads.

ground behind high, wide dunes or within the forested portions of the island interior are at the lowest risk for damage. Building within the forest reduces the amount of protection and also leads to increased degradation and killing of the exposed portions of the maritime forest from salt spray, wind, and wind-borne sand abrasion.

Risk assessment involves a two-step process, combining information on physical hazards with information about vulnerability (NRC, 1994). The needed information on physical hazards includes types of processes, their intensity and frequency, plus identification of areas of potential impacts. The vulnerability of a community refers to exposed population, the type and location of structures, critical facilities such as hospitals, shelters, and power plants, evacuation routes, and natural resources (NRC, 1994).

Defining the natural hazards, measuring the vulnerability, and evaluating the risks are the preliminary steps leading to avoiding the hazards and/or reducing the risks and in turn the vulnerability. *Mitigation* is the common term used by planners and managers for reducing the impacts and potential for loss of life and property from hazards. Although mitigation techniques are being applied in the coastal zone, property losses continue to climb.

Given that the coastal population growth will continue, coastal communities and society at large must take a series of collective actions to stem these catastrophic losses. The following are essential:

—Recognize that the most heavily developed coasts are barrier islands; fragile sys-

1.7 Little of this Pensacola, FL, home was left standing after Hurricane Opal passed over on October 4, 1995.
1.8 Only time will tell if the hard lessons of Hurricane Opal have been learned in areas such as Mexico Beach, FL. Photo by Craig Webb.

tems that behave differently from mainland coasts.
—Learn the "rules of the sea" as a basis for living with and managing barrier islands as well as for applying mitigation techniques.
—Develop new whole-island, "natural" mitigation techniques (i.e., islandwide approaches that mimic nature in preserving, augmenting, and restoring environments).

The 1995 hurricane season provided the perfect opportunity for the panhandle counties of Florida to implement the rules of the sea. Hurricanes Allison, Erin, and Opal (figs. 1.7 and 1.8) generously illustrated the lessons to be learned.

As we said in the preface, we do not recommend living on barrier islands and we definitely would not want our loved ones to live there. The hazards are numerous and difficult to avoid by evacuation. Development on the islands is destroying a critical and limited ecosystem. Read on, however, if you are willing to take your chances: this book will help you reduce your risk. We want to convince you to live by the rules of the sea.

Property damage mitigation in the coastal zone must be guided by coastal type, which is a function of geologic and oceanographic setting (processes, materials, landforms), as well as climate (processes, vegetation). Stability, elevation above flood level, and protective vegetative cover are commonly associated with low-risk mainland coasts. Rocky coasts and cliffs in stable, consolidated rock not subject to landslide processes or seismic instability, and at high elevations, are less likely to experience storm-surge flooding and direct storm wave-attack loss than low-lying, unconsolidated shores. The latter includes bluffs, plains, dunes, and other coasts developed in erodible sediments. Most would agree that it is foolhardy to develop a subsiding delta or mangrove swamp. Those same persons, however, regularly purchase and develop property on barrier islands (fig. 2.1), the most common and one of the least stable of coastal types on the U.S. eastern seaboard and Gulf of Mexico coast.

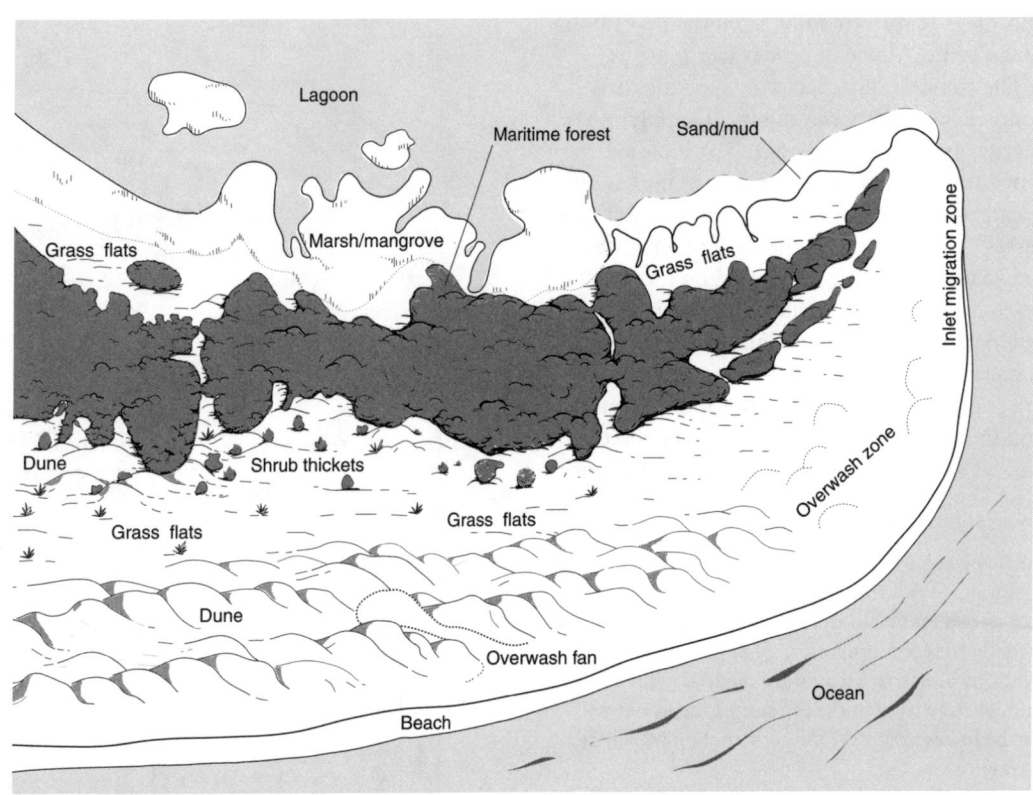

Barrier Islands: Built-In Hazards

The most dynamic of coastal types are the low-elevation sandy barriers that typically develop along coastal plains and deltas or as detached sandbars (spits) off headlands. The high-risk barrier islands of the United States are the primary geomorphic features of the U.S. Atlantic and Gulf of Mexico coasts, forming an almost

2.1 A typical undeveloped barrier island. This "ideal," here termed Pandora's Island, will be used as a model throughout the text. No two barrier islands are the same, but Pandora's Island has most of the components of typical temperate zone barrier islands.

unbroken chain extending from the tip of Long Island to the Mexican border (see fig. 1.2).

The coastal landscape is an evolving, dynamic balance between sediment supply, wave energy, and sea-level change. The building blocks of barrier islands are woven into an intricate and ever-changing pattern. Coastal processes act on materials, the stuff of which barrier islands are built. Beaches, dunes, island ridges, terraces, tidal deltas, marsh platforms, mudflats—every barrier island landform—are constructed of sediment, usually sand, shell debris, and mud. Add plant cover to the formula and the sediment can be trapped, anchored, and allowed to build up as grasses build dunes

2.2 Typical barrier island cross section (from Godfrey, 1976). Environments change quickly from end to end and from side to side. Each responds differently to storm processes. High-elevation, forested portions of barrier islands are the least vulnerable and when located in the interior of the island will be the lowest risk sites for construction (moderate to low risk).

2.3 Coastal environments and associated processes as natural hazards. The wider the bar, the more frequent and/or intense the process. Note that the environments subject to development all experience intense processes but generally have limited natural protection and are, therefore, hazardous. The exception is the so-called mainland forest, which is an island forest sufficiently protected from wind and salt spray to have nearly the same floral components as a forest well inland.

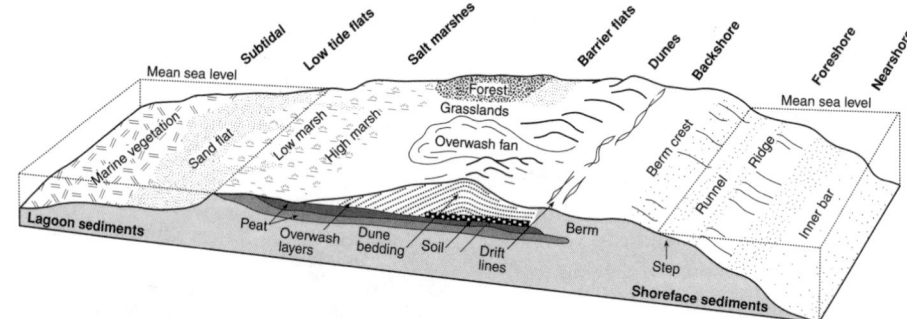

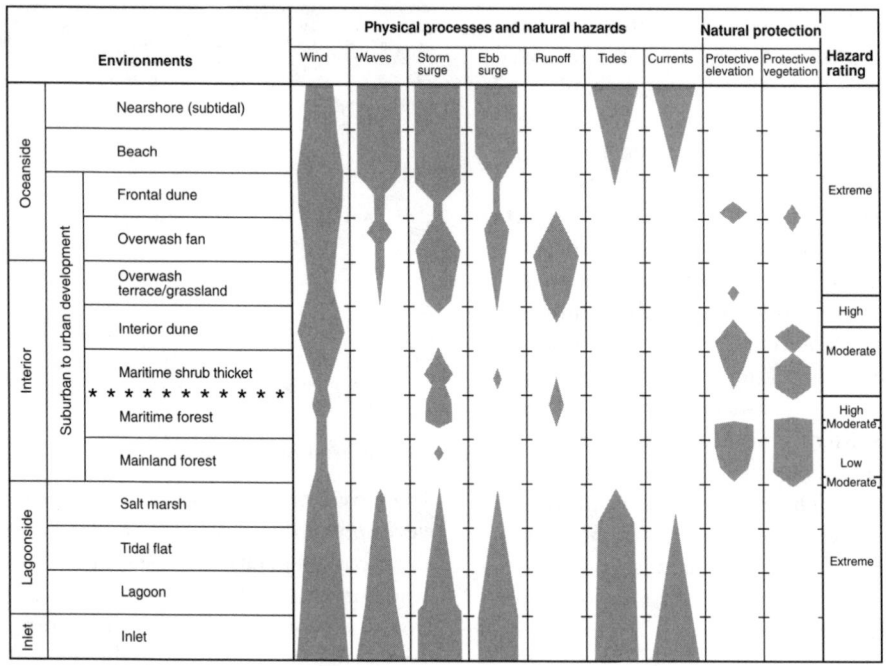

Environments		Physical processes and natural hazards							Natural protection		Hazard rating
		Wind	Waves	Storm surge	Ebb surge	Runoff	Tides	Currents	Protective elevation	Protective vegetation	
Oceanside	Nearshore (subtidal)										Extreme
	Beach										
	Frontal dune										
Interior	Overwash fan										
	Overwash terrace/grassland										High
	Interior dune										Moderate
	Maritime shrub thicket										
	********** Maritime forest										High [Moderate]
	Mainland forest										Low [Moderate]
Lagoonside	Salt marsh										Extreme
	Tidal flat										
	Lagoon										
Inlet	Inlet										

(left side label: Suburban to urban development)

* Occasional freshwater ponds

and marshes; grasses give way to shrub thicket and then to forest in the plant succession (fig. 2.2). The same basic environments are also found in many nonbarrier coastal locations. An idea of the physical processes active in each environment, their intensity and frequency, and the resulting hazard ratings are shown in figure 2.3.

Barrier islands act as the interface between ocean and land, between terra firma and the relentless power of the sea. Barrier islands bear the full impact of atmospheric and oceanographic energy, including winter storms, hurricanes, storm-surge flooding, and waves. The global sea-level rise is also adding to the instability of these islands, inducing landward migration. Barrier islands are unconsolidated masses of gravel, sand, and mud, surrounded by ocean and sound waters and characterized by low elevation, narrow width, and fragile vegetation cover. As a result, such islands are highly susceptible to wave erosion, overwash, longshore drift, flooding, flood scour, wind damage, and dramatic sand movement during storms. These hazards are often exacerbated by human alterations to the system such as construction of jetties and bulkheads, dredging of channels and finger canals, flattening of dunes, removal of protective vegetation, and unwise siting of buildings, roads, and utilities (fig. 2.4). When the processes are ignored and natural protection removed, the vulnerability to hazards is increased (fig. 2.5).

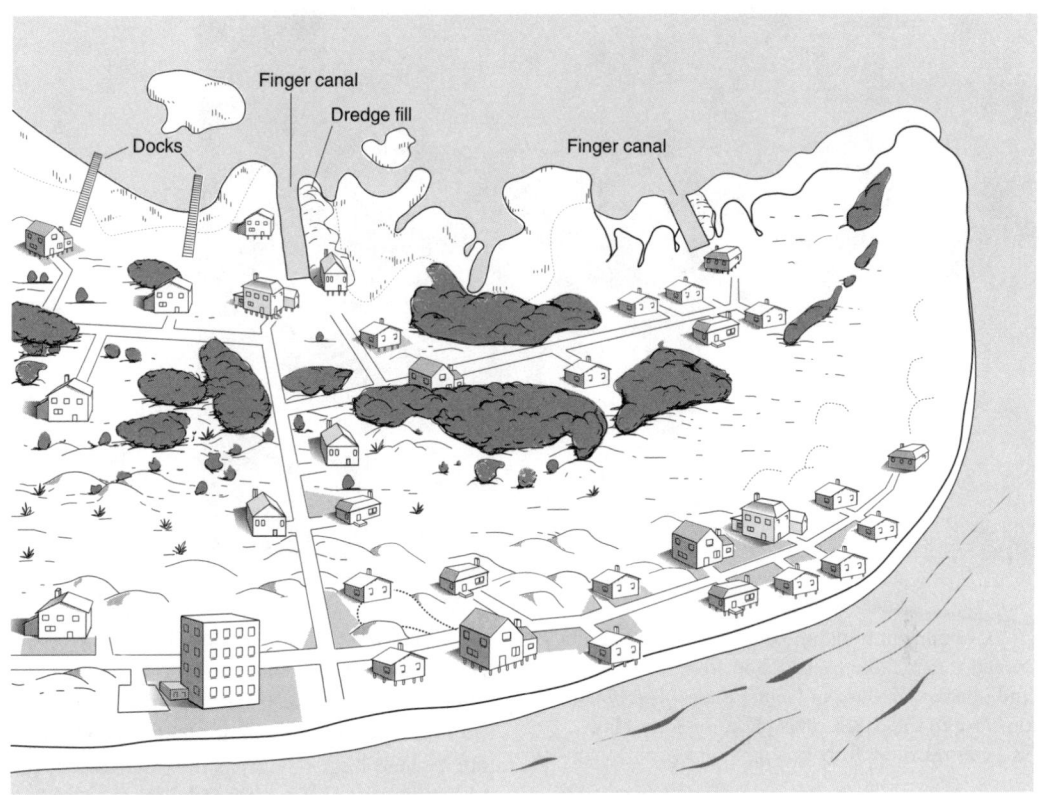

2.4 Development comes to Pandora's Island. The forest is partly removed. Dunes are flattened to form building sites. Finger canals are cut on the lagoon side and flat, straight roads are constructed through the dunes. Compare with figure 2.1.

Storms are the most dramatic of island hazards and receive most of our attention. Each big storm brings property damage and changes in the coastal landscape, and sometimes causes loss of life. During the past 100 years, an average of two major hurricanes have crossed the U.S. coastline every three years (Nuemann et

2.5 Development in danger during the 1991 Halloween storm. Note overwash in streets, flooded island interior, absence of frontal dunes, and houses too close to the beach. This photo of Kitty Hawk, NC, was taken by Rob Young.

al., 1989). Typically, two or three dozen winter storm systems (northeasters) impact the U.S. mid-Atlantic coast each year (Davis and Dolan, 1993), and similar storms affect the western U.S. coast.

The point is that living on a barrier island is like living on an active volcano: very scenic and very risky. You never know when it's go-ing to blow. We do know how barrier islands formed, how they evolve, and how they respond to changes, both natural and artificial. Shouldn't we apply this knowledge as we locate in these high-risk areas in order to reduce the loss potential from hazards?

Playing by the Rules

Nature is not subtle in trying to teach us of the dangers of coastal living. Every hurricane and northeaster (Ash Wednesday, 1962; Camille, 1969; Lincoln's Birthday, 1972; Gilbert, 1988; Hugo, 1989; Halloween, 1991; Andrew, 1992; the Storm of the Century, 1993; Emily, 1993; Opal, 1995) is a lesson that we should be living by the rules of the sea. We must recognize the principles from these lessons and put the rules into practice.

The need to establish a means of effectively and routinely evaluating (1) the potential risks from coastal processes to development, (2) disaster preparedness, and (3) disaster response (e.g., evacuation) becomes increasingly important. Deaths from coastal storms have decreased significantly over the past decades, thanks mostly to improved prediction and monitoring of storms and better evacuation and sheltering; but the potential for increased death tolls is very real. The costs from property damage are out of control (fig. 2.6). Reducing property damage is a national imperative.

The following chapters bring to light some new approaches to both coastal risk assessment and property damage mitigation that reverse the growing disregard for the important protective role of island environments and, where possible, suggest the enhancement of the natural protective features of islands. The fundamental first step for a community seeking to reduce property damage potential is to determine and map the risks. A simple, qualitative approach to risk mapping and evaluating the relative importance of coastal hazards is presented. Mapping is at the reconnaissance level, defining broad hazard zones within coastal communities based on geologic and oceanographic setting. The principles presented apply to most Atlantic and Gulf coast barrier islands. Many of the examples presented are drawn from the Carolinas, the focus of our specific pilot and demonstration studies, but any barrier island community can apply this mapping approach. The first step for an individual buying property is to evaluate the specific homesite using the same parameters used in risk mapping. In the past, if such an evaluation was considered at all, it was the first and last step. Experience calls for broader risk assessment at the community and islandwide levels.

The objective is to promote a new attitude about living in the coastal zone, particularly on barrier islands. The ultimate goal of property damage mitigation can be viewed as a four-step process:

(1) understanding/recognizing hazards and vulnerability

(2) mapping zones of risk

(3) developing site-specific mitigation techniques, including preservation, augmentation, and restoration of natural environments (PAR)

(4) implementation of mitigation recommendations.

Island Risk Assessment

Very large differences exist in property damage risk exposure between different islands, between different segments of the same island, between the front and back sides of an island (see fig. 2.3), and even between adjacent buildings. Thus a relative level of risk can be determined for a given structure, a property, a community, or an entire island with respect to the potential for damage from hurricanes, other coastal storms, and ongoing erosion.

Our approach to risk assessment is from a geologically based understanding of coastal processes and from observations of impacts of storms on coastal development. Coastal hazards, from a geologic point of view, include both the intense, short-duration physical processes associated with storms as well as the intermediate-term cumulative effects of such events (e.g., continuous coastal erosion and shoreline encroachment). Longer-term processes (e.g., subsidence of a delta) also may be important. Storms come in all shapes and sizes, from modest tropical depressions to category 5 hurricanes, from passing cold fronts to immense winter storms (northeasters). Table 2.1 gives an idea of the frequency, intensity, and duration of several types of natural hazards that impact coastal areas and compares their overall severity and impact with some other common natural hazards that have affected the United States in recent years. Obviously the recognition and mitigation of coastal hazards are important concerns facing our nation as we prepare for the twenty-first century.

The assessment technique and property damage mitigation recommendations presented here are from a conservative perspective. That is, the following two major considerations are the basis of the text's recommendations:

—Winds of 120 miles per hour (190 kilometers per hour), essentially the eye wall of a category 3 hurricane or the more distant portions of a stronger hurricane, is the *maximum* condition; our evaluations are too low for category 4 and 5 hurricanes or their equivalent.

—Housing sites recommended are those we would feel comfortable recommending to our parents or grandparents (assuming they were imprudent enough to want to build on a barrier island).

2.6 The historic record of the decrease in deaths and increase in property damage from hurricanes in the United States. A better storm warning system explains the former and denser development explains the latter (information from Williams, 1992).

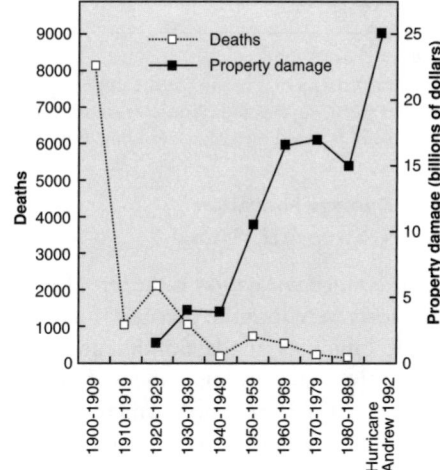

Table 2.1   Ranking of Great Natural Hazards

Overall Rank[2]	Event	Grading of Characteristics and Impacts[1]								
		Degree of Severity	Length of Event	Total Areal Extent	Total Loss of Life	Total Economic Loss	Social Effect	Long-Term Impact	Suddenness	Occurrence of Associated Hazards
1	Drought	1	1	1	1	1	1	1	4	3
2	Tropical Cyclone*	1	2	2	2	2	2	1	5	1
3	Regional Flood	2	2	2	1	1	1	2	4	3
4	Earthquake	1	5	1	2	1	1	2	3	3
5	Volcano	1	4	4	2	2	2	1	3	1
6	Extratropical Storm*	1	3	2	2	2	2	2	5	3
7	Tsunami*	2	4	1	2	2	2	3	4	5
8	Sea-Level Rise*	5	1	1	5	3	5	1	5	4
9	Landslides	4	2	2	4	4	4	5	2	5
10	Beach Erosion*	5	2	2	5	4	4	4	2	5
11	Tornado	2	5	3	4	4	4	4	2	5
12	Blizzard	4	3	4	4	4	4	5	1	5
13	Ocean Waves*	4	4	2	4	4	5	5	3	5
14	Localized Strong Wind*	5	4	3	5	5	5	5	1	5
15	Subsidence*	4	3	5	5	4	4	5	3	5

* Hazards of coastal zone.

[1] = The characteristics of a hazard or the importance of its impacts are ranked on a scale of 1 (largest or greatest) to 5 (smallest or least significant).

[2] = Overall rank is based on average grading. Source: Adapted from Bryant, 1991.

Property Damage Mitigation: Regulatory, Structural, Natural

Property damage mitigation is the term for collective efforts to reduce the potential for storm damage to buildings, roads, utilities, and infrastructure. The goal is to offset some of the danger in the rush to the shore which has resulted in more and more property and people at risk. The current sea-level rise, plus predictions of both an acceleration in that rise and an increased frequency and intensity of storms, add greatly to the urgency of taking measures to assess coastal hazards and to reduce property vulnerability.

Observations of several barrier island communities after Hurricanes Gilbert (1988) and Hugo (1989) indicate that property damage potential can be lessened by prudent site selection and location of buildings. Likewise, post-storm observations also indicate that reliable means exist by which property damage can be mitigated (e.g., dune construction and road reorientation). The recent passages of Hurricanes Gilbert, Hugo, Bob (1991), Andrew (1992), Iniki (1992), Emily (1993), and Opal (1995) have demonstrated the need for a new approach to mitigation in order to stem the tide of ever-increasing property damage from storms.

Past approaches to property damage mitigation are discussed by Godschalk, Brower, and Beatley (1989) and include such broad categories as

—engineering to armor the coast
—engineering to strengthen buildings
—land-use planning to avoid construction in hazard areas

These mitigation techniques are primarily applied through regulations (e.g., setback requirements, building codes, permit requirements) and heavy reliance on engineering (structural solutions).

Property damage mitigation includes specific goals such as

—reducing damage to structures
—preserving natural environments
—increasing evacuation capacity
—locating new structures out of hazardous areas
—relocating existing structures out of hazardous areas
—providing safe storm shelters
—structurally altering the environment

For the most part, approaches to property damage mitigation developed in recent decades focused on the regulatory and structural methods, efforts that may be misplaced. Traditional mitigation approaches are somewhat myopic, focusing mainly on the beach and island front, although some attention has been given to protecting critical environments (e.g., marshes).

A new, broader approach, viewing the entire barrier island as well as the adjacent nearshore ocean floor, is needed, mitigation that is based on the entire complex system and applied islandwide. To decrease the risk to lives and property on barrier islands, let's apply the rules of the sea. It is hoped this new approach can be used as a tool by coastal planners and managers in making their mitigation plans and post-storm reconstruction guidelines. For island dwellers, understanding this approach should provide a basis for wiser coastal property purchases no matter where on an island the house or lot is located. And existing communities can identify problem areas where environmental features can be repaired or enhanced to restore natural mitigation.

The natural coastal environment is quite capable of absorbing the impact of a coastal storm, albeit with some change in the coastal geomorphology. Changes include raising the island elevation by dune formation or deposition of storm overwash sand; reducing the island elevation by destruction of dunes; changing island width and shape by shoreline retreat on either the ocean or sound side or shoreline advance (beach accretion on either the ocean or sound side); and steepening or flattening the beach profile. Such changes may be permanent or temporary depending on the coastal environment in question and its unique geologic, oceanographic, and climatic setting. On barrier islands, such changes contribute to the natural long-term processes of barrier island migration. How and to what degree development interferes with these natural coastal changes will determine what steps can be taken in a given area to reduce the potential for damage from future storms. The rule is that mitigation should mimic nature.

PAR: Preserve, Augment, Restore

The guide to property damage mitigation presented here is based mainly on experience with past storms. Observations were made after several hurricanes, tropical storms, and winter northeasters along the southeastern U.S. coast and the coasts of Mexico and Puerto Rico (Bush, 1991, 1994; Hall et al., 1990; Gayes, 1991; Lennon, 1991; Priddy, 1991; Thieler, Bush, and Pilkey, 1989; Thieler and Bush, 1991; Thieler and Young, 1991; Bush and Pilkey, 1994), as well as the Gulf of Mexico (Penland, Nummedal, and Schramm, 1980; Kahn, 1986; Morton et al., 1985). These studies of hurricane damage indicate that, in most cases, useful and economically reasonable steps could have been taken before these storms that would have reduced property damage significantly. These mitigation actions range from preserving natural areas that afford protection (e.g., forest cover) to augmenting such environments or landforms (e.g., frontal dunes) to restoring or repairing such features when altered or damaged (e.g., interior dunes).

Methods of protecting buildings and other property from coastal storms and erosion range from simply planting dune grass and placing sand-trapping fences for encouraging dune growth to designing sand emplacement and rebuilding of beaches, roads, and service systems —from efforts costing very little to projects costing millions of dollars per mile of island length.

The process-oriented perspective (i.e., to mimic nature) presented here provides for identification of hazard areas based on likely processes such as potential overwash zones and potential inlet formation areas; zones of inlet expansion or migration; and identification of potential flooding problems and flood zones. The interrelationship of processes, materials, and vegetation is considered. Moreover, a geologic perspective allows special insights into repairing damage already done to the natural environment, for example, restoring beaches, rebuilding excavated interior dunes, plugging dune gaps, reestablishing destroyed maritime forest, curving roads around natural island topography, and putting roads and walkways over rather than through such topographic features. In addition to repairing damage done to the island, steps can be taken to enhance the natural protective capabilities of coastal environments, such as planting marsh grass to slow lagoonside erosion, replenishing beaches, constructing dunes, and encouraging maritime forest growth. Planning can avoid hazard enhancement by not creating overwash passes, not interrupting sediment supply, not disrupting stabilizing ground cover, and not creating potential for new inlet formation.

The Rules of the Sea

Recognition of the physical processes active within coastal environments is the fundamental step toward recognition of hazard areas and forms the basis of a "coastal processes approach" to property damage mitigation. The following points are the central tenets of *Living by the Rules of the Sea*:

(1) *The coastal zone is unique and requires unique management strategies.* Barrier island environments are far more dynamic than mainland areas. The traditional grid-development pattern and related construction used on the adjacent coastal plain and the stable interior of the continent is inappropriate for a barrier island and *increases* the probability of impact by natural processes; that is, the type of development can often increase the risk from natural hazards.

(2) *Coastal physical processes must be identified and understood from a whole-island perspective.* Island, marsh, dune, beach, and offshore are all part of one large interrelated geobiological system. Building in the path of the processes creates the hazard.

(3) *Property damage potential is site-specific and each site is different.* Each area presents a unique set of circumstances that require unique solutions; however, broad general principles can be drawn on to develop these solutions.

(4) *Property damage mitigation must be from a whole-island perspective.* Barrier islands behave as units or systems in terms of process response. Recognize that mitigation can no longer be considered something for only the first one or two rows of houses. Rather, the whole island and complete processes/materials system must be considered.

(5) *Relative risk areas can be recognized on the basis of well-defined criteria.* By observing these criteria, development can be directed away from inlet hazard areas, potential overwash zones, low-elevation areas, and so forth. If location out of harm's way is impossible, then mimic the process; for example, allow overwash to occur, dunes and beaches to migrate, forests to grow.

(6) *All coastal hazard evaluation and mitigation must consider a rising sea level.* The present interglacial period is resulting in a worldwide shoreline migration as sea level rises over a sloping land surface. The sea-level rise is likely to continue in the foreseeable future and will likely even accelerate over the next 50 to 100 years due to the greenhouse effect. Solutions must take into account barrier islands continuing to move landward as the sea level rises. The storm-to-storm crisis approach should be replaced with a search for long-term solutions for a long-term problem.

(7) *Repair alterations due to development.*
Damage to the natural setting reduces the
natural protective qualities of the island. Such
damage must be repaired in order to mitigate
future property damage. In many cases such effort will entail little more than restoring relatively small areas to the pristine state of
predevelopment by rebuilding dunes, planting
grasses, or replacing maritime vegetation. In
some cases, the effort will require community
spirit, political grit, and adequate financing
(e.g., beach replenishment, relocation of roads
and services). *Augmentation* of such environments should be given the same emphasis as
repair.

(8) *Conserve sand.* Sand volume must be
maintained or increased. Emplacing new sand
from an off-island source is better than moving
sand from place to place on an island. Sand,
like water and air, is a resource to be conserved.

(9) *Conserve vegetation cover.* Vegetation is
the closest thing to a sign of stability on a barrier island. Vegetation traps and anchors sediment whether it be dune grass building dunes,
forest canopy protecting undergrowth, or
marsh grass trapping sand and mud and building flats in the lagoon or sound. Removal of
vegetation destabilizes the land, raising the
hazard potential.

(10) *Conserve landforms.* This rule follows
from those that go before it. The landform
(e.g., dune, island terrace, tidal flat, beach) is a
sedimentary body that formed in response to
processes, sediment supply, and vegetation
cover. When the landform is altered, the stability is altered. Stabilization may be augmented by adding and anchoring sediment,
planting natural species, or even constructing
an artificial landform.

Hazard assessments, vulnerability determi-
nations, risk estimates, and mitigation recommendations are the realm of scientists and engineers. Implementation, however, falls into
the political/legal arena. For example, dune
protection ordinances, community mitigation
plans, and zoning must be based on scientific
principles. Their enforcement and the
financing to build dunes, move houses off the
beach, and preserve wetlands must be carried
out by elected and appointed officials. The
rules outlined above provide the basic principles that should underlie and guide such
implementation. The impetus to make such
mitigation happen must come from individual
property owners, whether the motive be continued enjoyment of living in the coastal zone
or continued profit from recreational and
business property investments.

3   Storm Processes as Coastal Hazards

Once the coastal zone, especially barrier islands, is recognized as unique, the next rule in assessing coastal hazards to reduce property damage is identifying the natural storm forces or physical processes that result in environmental impact and potential property damage. Storm processes, acting singly or in various combinations, are the destructive forces, the hazards, of concern (fig. 3.1): these are wind, waves, coastal and inlet currents, storm-surge flooding, and storm-surge flood and ebb currents. Wind, waves, and rising water receive a lot of attention during hurricanes and, in fact, account for most of the damage. Currents are responsible for moving vast amounts of sediment during storms. The onshore movement of water, the storm surge, causes flooding and may induce scouring currents around and behind structures. The rising water level allows the zone of wave attack to move inland and sediment to wash over onto the land. Storm-surge ebb, or the seaward return of storm surge, is a less familiar storm process that may erode new inlets and contribute to the overall erosional damage. Table 3.1 and figure 3.2 summarize the various storm processes and their effects and provide a basis for defining areas in which risks are likely (see chapter 2, rule 5).

Natural Processes: Energy in Motion

Storm processes rarely act separately. That is, wind, waves, and currents are all active at the

Table 3.1   Storm Processes and Their Effects

Storm Wind:
Direct wind attack on buildings
Flying debris (missiling)
Sand onto/off island (burial or erosion)
Vegetation loss (blow down, salt spray kills, sandblasting of leaves)

Storm Waves*:
Direct wave attack on buildings
Floating debris (ramrodding) from buildings and attachments
Scouring around foundation footings
Shoreline retreat on lagoon shore (erosion)
Shoreline retreat on ocean shore (backbeach erosion)
Overwash (burial and blockage)
Dune loss
Scarping of fastland (nondune)
Vegetation loss (erosion, saltwater kills)
Local flooding
Strong longshore currents, remove sand from area

Storm Surge*:
Flooding
Floating debris (rafting)
Lagoon shore retreat
Ocean shore retreat
Widening inlets
Changing channel locations in inlets
New inlet formation

Increases zone of wave influence (elevates waves, moves waves landward)
Overwash (burial and blockage)
Scouring of cross-island channels and undermining of structures
Scouring around foundation footings
Vegetation kills (saltwater kills, including saline groundwater contamination)
Saltwater flooding impacts (sterile soil, contaminated groundwater
Drives offshore-directed currents, permanently removing sand from system

Storm-Surge Ebb:
Widening inlets
Changing channel locations in inlets
Formation of new inlets
Scouring of cross-island channels
Scouring of offshore channels
Scouring around foundation footings and other hard structures
Emplacement of debris offshore (swimming and boating hazard)
Sand removal/permanent sand loss

High Rainfall:
Water damage to buildings when coupled with high winds
Enhanced flooding
Enhanced erosion due to runoff

*Effects are enhanced and extend farther into the interior of the island if the storm makes its landfall on a high tide, especially a spring high tide.

Soundside flood and erosion zone

Inlet migration zone

Overwash penetration and potential for new inlet formation

Interior flooding along roads and low areas

Oceanside flood and erosion zone

3.1 Zones likely to be affected by various storm processes on Pandora's Island. Each island is different and is in a different oceanographic setting. Therefore, to understand hazards to property on islands, it is necessary to study past storms and their effects on island environments.

same time and combine to form secondary processes. For example, storm surge is formed by several processes acting together, any one of which may be dominant during any given storm or for a given period during a certain storm: wind pushes water toward shore, waves push water toward shore, low pressure allows doming of the sea surface, and the rotating

winds of a hurricane actually cause the water surface near shore to spiral higher as the surge moves into the shallower water. In the following section, we look at storm processes individually to better understand their actions during storms.

Wind

The most common and, often, the most costly of storm hazards causing damage to buildings is direct wind impact on structures, including flying debris (known as missiling). In addition, strong winds can destroy vegetation by uprooting and knocking over trees, defoliating trees and other vegetation, blowing down shrubs and grasses, and damaging leaves directly either by blasting them with airborne sand or carrying damaging salt spray inland (fig. 3.3). The same salt-spray pruning effect that produces the sloping, sculpted profile of maritime forest and shrub will kill or damage inland vegetation that is not salt tolerant. Strong winds can also be responsible for transporting sediment onto and off an island (NWS, 1993).

Storm Waves

Property is damaged by direct wave attack on structures or by pummeling structures with floating debris, a process called ramrodding (fig. 3.4). Probably the only type of building capable of surviving direct wave assault un-

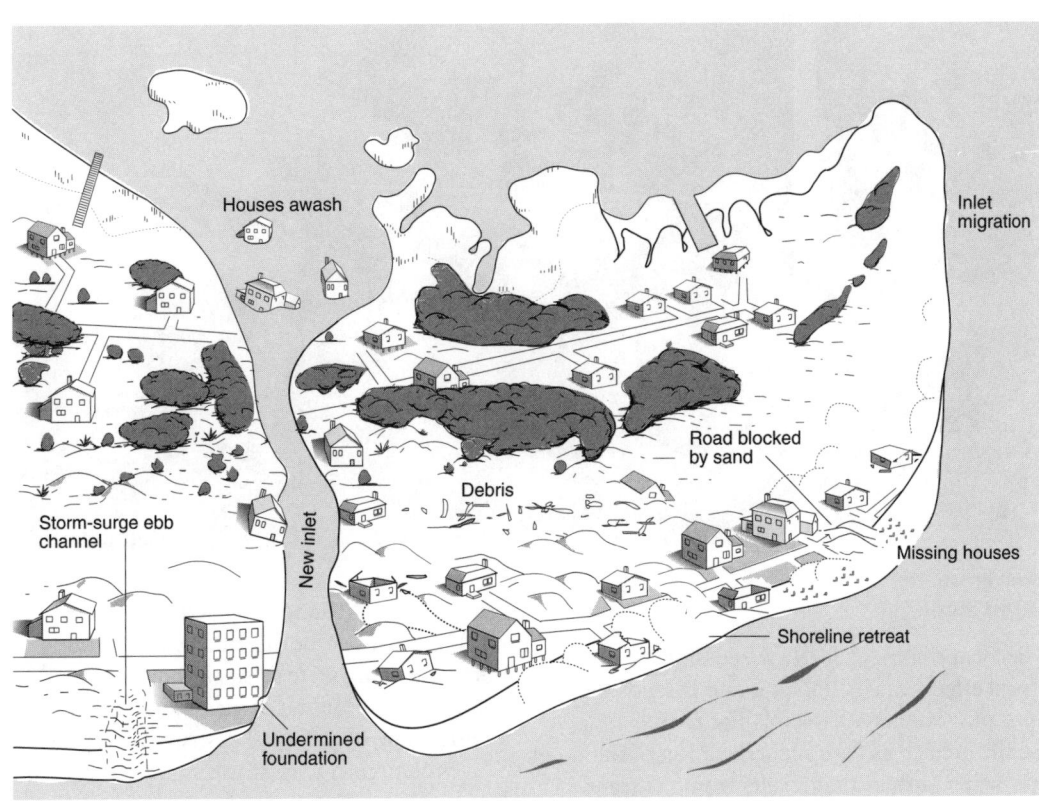

3.2 Pandora's Island after the big one. A new inlet that formed followed the combined path of a road and finger canal. The shoreline has retreated and some houses have disappeared.

scathed is the concrete pillbox. Even lighthouses have gone down under wave attack. Waves are also responsible for shoreline erosion (on both lagoon shores and ocean shores) as well as dune erosion, overwash, and destruction of vegetation.

3.3 Hurricane Hugo had much impact on vegetation. (above) In Puerto Rico the trees in the foreground were completely stripped of leaves. (below) In South Carolina's Francis Marion National Forest, pines covering thousands of acres were snapped off 10 to 20 feet (3 to 7 meters) from their base.

3.4 Damage caused by the waves from Hurricane Gilbert (1988) striking concrete buildings on the Yucatán Penninsula, Mexico. There is essentially no way to defend a building against direct wave attack, except perhaps construction of a concrete pillbox!

3.5 Storm surges may float one house into another, as in this case in Garden City, SC, after Hurricane Hugo. Flying and floating debris is a major cause of damage in hurricanes. Photo by Rob Thieler.

Storm Surge

The local rise in sea level caused directly by storm effects causes flooding and extends the zone of wave impact inland. The term is technically defined as "the superelevation of the still-water surface that results from the transport and circulation of water induced by wind stresses and pressure gradients in an atmospheric storm" (Simpson and Riehl, 1981). *Pressure gradient* refers to the lowered atmospheric pressure in storms, which by itself can cause a rise in sea level. Storm-surge impacts include flooding, floating structures off their foundations, and floating debris inland, sometimes with ramrod force (fig. 3.5). The initial flow over and around obstructions (e.g., pilings) may cause scouring and sediment transport. The rising water also elevates waves and increases their landward incursion, resulting in a wider zone of potential destructive impact. Waves combined with storm surge act to wash beach sand onto the island, forming "overwash" deposits. Saltwater flooding kills or damages inland plants.

Currents

Storm-generated currents transport water, sediment, and storm debris both parallel and perpendicular to the coast. By far the greatest volume of water moved is in an alongshore direction, as storm waves approaching the shore at an angle set up a current in the general direction of wave travel. Because storm waves are so large, they begin to break much farther from shore than normal or fairweather waves. The net effect is a widening of the surf zone. The *longshore current* described above is a *surf-zone phenomenon*, so merely by enlarging the width of the surf zone, more water is moved parallel to shore. This current can move sediment (along with trees, sand fencing, dune-crossover stairways, decks, and other building debris) out of one local area and into another, resulting in a loss of sediment from one portion of the coast. This loss may be temporary or permanent, depending on many other factors. In some cases *rip currents* may be intensified during part of the storm, making conditions even more dangerous for those foolhardy enough to try to surf or swim during a storm.

Storms are also responsible for *bottom currents,* which move at high angles, sometimes perpendicular, to the shoreline. These currents are set up because of the local rise in sea level caused by the storm surge. The storm surge sets up a sea surface that is actually tilted away from land. The water above normal sea level tends to "flow downhill" back out to sea. These types of currents may not be very strong, and they certainly move a much smaller volume of water than longshore currents, but they can be responsible for moving great quantities of sediment away from the

beach and into deeper water. They are efficient at moving sediment because they are accompanied, and even partly created, by large storm waves. Storm waves, because of their great turbulence, are very efficient at resuspending sediment off the sea bottom and carving up the beach and dunes. Once sediment is moved up off the sea bottom and put into the water column, only the slightest current is needed to carry the sediment away. These seaward-flowing currents can actually carry sediment many miles offshore, often so far that the sand is lost from the beach system. This loss is true erosion of the shoreline: a permanent net loss of sediment from the beach and dunes.

3.6 This storm-surge ebb channel on Folly Island, South Carolina resulted in the destruction of the seawall. Damage done by storm-surge ebb is often incorrectly attributed to the initial storm surge crossing the island. Photo by Rodney Priddy.

Currents and patterns of flow (channel positions) in inlets may be modified by the combined effects of storm surge, stage of the tidal cycle, increased rainfall runoff, and storm-surge ebb scour (see below). Changes in channel positions during storms may cause erosion of or deposition on adjacent islands. Murrells Inlet, South Carolina, for example, has historically migrated to the south with periodic updrift relocation during storms. Others, such as Bogue Inlet, North Carolina, or Capers Inlet, South Carolina, demonstrate relocation of the main inlet channel position within the tidal delta, not relocation of the entire inlet. Chapter 7 discusses inlet dynamics in greater detail.

Storm-Surge Ebb

The "piled up" storm-surge water flows back to sea, either by the force of gravity alone or when driven by offshore-blowing winds, generating an erosive ebb current. This type of current is different from the ones described above because it occurs while the storm is moving out of the area or diminishing. Storm-surge ebb can cause an existing inlet to change shape; can create a new inlet, such as on Pawleys Island, South Carolina, during Hugo; can scour shallow cross-island channels; can transport storm debris (including houses) offshore; and can cause permanent removal of sand from the beach/dune system to the deeper offshore (fig. 3.6). After Hugo, the shoreface in front of Myrtle Beach, South Carolina, was covered with deep scour tracks, perpendicular to the shoreline. These channels were eroded by the storm-surge ebb.

Human Modification of the Coast

Construction in the coastal zone may enhance or otherwise alter the natural processes and their resulting impacts. Roads and beach access paths perpendicular to the shore that penetrate the dune line may become overwash passes or focal points for storm-surge flood or ebb currents. Seawalls may redistribute wave energy or obstruct sediment movement. Jetties may block great volumes of sand from being transported along the coast, resulting in deposition of sand and beach widening on the updrift side and a long-term sand deficit and erosion on the downdrift side. Ground-level houses and closed-in ground floors of houses on stilts may obstruct the passage of overwash sand, which is then lost to front-side erosion. Where vegetation cover has been removed, erosion by wind or water may occur.

Sediment Supply: No Deposit–No Return

Sand is the barrier island's lifeline between sea and land. Wide sand beaches buffer storm waves. Sand dunes are stores that waves draw on in the big event, and as such are the last line of natural protection between the ocean and island property. Sand fill of former inlets makes up the barrier island's base, and the sand of

overwash and dunes is the cap of the island standing above sea level. Put simply, under natural conditions a barrier island with a plentiful sand supply will be a healthy, robust island. But it will not be a stationary island if sea level is rising. Nature rearranges the sediment and landforms continuously in response to changes in wind, waves, currents, and sea level, whether short term (e.g., storm surge, tides) or long term (e.g., climate-controlled rise and fall of the sea). Use of the term *nourishment* to describe the artificial building or maintenance of a beach is an accurate reflection that beaches, dunes, in fact the entire barrier island, need a sand supply to survive. The same need exists for mud, fine-grained sediment, on the back sides of islands to build marshes and associated features.

Community officials, planners, developers, and property owners in the coastal zone should know the answers to these questions:

—Where is our sediment coming from?
—What types of sediment are being supplied?
—What is the amount of the sediment supply?
—How does the supply vary?
—Where is the sediment going?

If you don't know the answers and proceed with construction, the chances are good that one or more processes will be modified, the sediment supply disrupted, and the property vulnerability increased. A century of beach erosion problems has taught us that shore-harden-

ing structures (e.g., groin fields, seawalls, breakwaters) cut off sand supply to downdrift beaches, causing them to erode (see chapter 5). The same principles apply across and along the entire barrier island.

The questions above are not just for beaches on the island front, but apply to all island environments. Protective dune lines on natural barrier islands re-form after hurricanes as the wind "banks a deposit" that waves will draw from on the next rainy day (i.e., northeaster or hurricane). The overwash carries sand to the interior and back side of the island, building its elevation or allowing it to migrate. When we densely develop islands, the natural, post-storm healing cannot take place. If we are to mitigate future property damage then we must attempt to restore landforms as nature would restore them. Replenishing beaches and rebuilding dune fields by planting dune grass or erecting sand fencing are obvious mitigation strategies, but they are techniques applied primarily at the front side of the island.

In the past (and sometimes in the present), great quantities of sand were removed from various parts of barrier island systems. Dredging of navigation channels removed the "spoil," often good quality sand, to offshore "disposal" sites. Migrating sand dunes were considered a nuisance and were flattened, and excess sand was removed from islands. Interior dunes (and maritime forests) were cleared, flattened, and removed for building sites (fig. 3.7). As part of the clean-up effort after

3.7 Area of home development on Emerald Isle, NC, where maritime forest has been cleared and high dunes leveled. An area of moderate risk has been transformed to one of extreme risk.

storms, overwash sand was cleared from the streets and hauled off the island. And tons of sand have been mined from barrier islands as a one-time construction commodity. Sometimes, as on Edisto Island, South Carolina, the island itself was mined to furnish sand for its beach replenishment.

The rule to *conserve sand* (see chapter 2, rule 8) must be applied islandwide, front-middle-back, inlet-to-inlet, and not just at selected sites, although conservation measures may range from islandwide projects down to how each individual property is developed and maintained. If sand is required to be put back into the system, it should be borrowed from natural "sinks." A sediment sink is analogous to the accumulation at the end of a conveyor belt—a place where sediment transport ends.

The sinks we would consider are mainly ancient types of sand deposits, not part of the active coastal system. Stranded barrier islands from ancient, higher stands of sea level might be one example. A river delta is another example of a sediment sink, but deltas are not present along many of the coasts where barrier islands form. Examples of exceptions include the Brazos River delta in Texas and the Santee River delta in South Carolina. Another type of sink is the sandy river channel from ancient river locations, which are numerous on most coastal plains and sometimes present in the older offshore sediments. Much of the sand used for replenishment projects in Myrtle Beach, South Carolina, comes from ancient river channels on shore. Offshore sand shoals and tidal deltas are looked at longingly as sources of sand for today's island sand needs. However, these sinks are still part of active systems, and their removal may have negative consequences, such as changing wave refraction patterns or tidal flow, which may create erosion problems somewhere else. Also, the effect on plant and animal populations is often difficult to predict, but is almost always negative.

Sediment supply of sand and mud is just as critical to the back sides of barrier islands as to beaches and interior dunes. Sand migration across the island and through the inlets provides a mechanism for landward migration of the entire island system. The resulting sand platforms (flood tidal deltas and tidal flats) provide the base for salt marsh growth. Salt marshes trap additional sediment and build up the back side of the island, protecting it from shoreline erosion.

Vegetation Cover: The Cap on the Deposit

If there were no vegetation on barrier islands, the sedimentary landforms would never be stable. The scene would be like a desert dune field: remolding, shifting, constantly changing. Nature's addition of plant cover to the processes and material formula provides mechanisms to trap and hold sediment in place. The result is that vegetation plays a significant role in barrier island dynamics.

Beach and dune grasses baffle wind currents so that sand is deposited, burying the grass. Some of these species are adapted to burial—it stimulates their growth, and more sand is trapped in successively higher, wider, and laterally continuous dunes. In a similar fashion, marsh grasses trap sediment, allowing the floor of the marsh to build upward. Roots of plants anchor the sediment. Plant succession from grasses to shrubs to trees creates a new environment and a protective cover. Fire, off-road vehicles, foot traffic, and, most often, construction remove the protection, allowing erosive processes to again dominate. Plantings of ornamental species ill adapted to the barrier island setting are of little value as protective plant cover. Often, dunes and their native vegetation cover are removed or flattened for planting of typical inland grass lawns. These grasses, ill suited to the harsh coastal environment and the very sandy soil, require large amounts of water to survive. When they don't

survive, sand is destabilized and a once flourishing and protective dune is reduced to an active sand flat, offering absolutely no protection from storms.

Again, the damage is not restricted to island fronts. Removal of marsh grass, channeling and causeway construction in marshes and sounds that result in plant kills, and mangrove removal (e.g., in Florida and the Caribbean) result in erosion on the back sides of islands. Opening clearings or breaks in the canopy of maritime forests results in loss of stabilizing undergrowth and wind protection, increased likelihood of tree blowdowns, and surface runoff or blowouts if on an old dune ridge.

The rule is: *Conserve plant cover*. Communities, developers, and property owners in the coastal zone should be able to answer these questions:

—What is the natural distribution of native vegetation?

—Where are the areas of maritime and upland forest?

—How is the vegetation cover being, or likely to be, disturbed by development?

—What can be done to protect, restore, or expand natural vegetation?

The Physical Nature of Hurricanes: All the Processes Rolled into One

Hurricanes are responsible for most of the storm-related coastal property damage in the United States; however, other types of storms,

particularly northeasters along the East Coast and southwesters on the Gulf coast, certainly are important. The actual processes that affect the coastal zone are similar in all storms, but they are most intense in hurricanes. During the relatively hurricane-free period from the 1960s until the 1989 brush by Hugo, the majority of today's coastal residents and property owners had not experienced the full force of such storms. (Historical hurricanes that have hit the United States are discussed by Hebert and Taylor [1988]). This led to a disregard of the hurricane menace and increased development in high-hazard zones. In rapid succession, Hugo, Bob, Andrew, Iniki, Opal, and near misses by Emily, Gordon, and Felix changed all that. The odds are evening out, and time is not on the side of coastal development.

Each year on June 1 the official hurricane season begins. For the next five or six months conditions favorable to hurricane formation can develop over the tropical to subtropical waters of the Western Hemisphere. Early-season tropical cyclones form mostly in the Gulf of Mexico or Caribbean Sea, where the waters can heat up faster than in the Atlantic Ocean. The monster hurricanes that strike the east and Gulf coasts of the United States usually originate later in the season (August, September, and October) in the eastern North Atlantic and grow on their long slow trek across the ocean.

Once formed, the hurricane mass begins to track into higher latitudes and may continue to grow in size and strength. The velocity of this tracking movement can vary from nearly stationary to greater than 60 miles per hour (about 100 kilometers per hour). When a hurricane makes landfall, the destructive forces are at their maximum in the area to the right of the forward motion of the eye, but the entire landfall area will experience the severity of the storm. The right-of-the-eye effect means that significant destruction can be generated even by a storm that passes offshore, particularly for a hurricane that tracks north to south, along the shoreline. One should not feel any security, however, in the knowledge that one is to the left of the eye! Even where the tidal range is small, if the hurricane comes on a high tide, especially a spring high tide (highest high tide), the effects of storm-surge flooding, waves, and overwash will be magnified. For more information on hurricanes, see Simpson and Riehl (1981), Hebert and Taylor (1988), Barnes (1995), and any of several natural hazards textbooks.

Hurricane Probability and Rank

The probability that a hurricane will make landfall at any given point along the coast in any one year is low, and the probability of a great hurricane makes such an event seem unlikely; but low probabilities give a false sense of security because the lessons of hurricane history tell us that in the lifetime of a structure such a storm is almost a certainty. Further-

Table 3.2   The Saffir/Simpson Hurricane Scale

Scale Number (Category)	1	2	3	4	5
Central Pressure:					
millibars	≥980	979–965	964–945	944–920	≤919
(inches of mercury)	(≥28.94)	(28.91–28.50)	(28.47–27.91)	(27.88–27.17)	(≤27.16)
Winds:					
mph	74–95	96–110	111–130	131–155	>155
[kph]	[119–153]	[154–177]	[179–209]	[211–249]	[>249]
(meters/ sec)	(32–42)	(42–49)	(50–57)	(58–68)	(>69)
Surge:					
feet	4–5	6–8	9–12	13–18	>18
(meters)	(1.2–1.5)	(1.8–2.4)	(2.7–3.7)	(4.0–5.5)	(>5.5)
Damage:	Minimal	Moderate	Extensive	Extreme	Catastrophic

Source: Developed by H. Saffir and R. H. Simpson (Simpson, 1974).

more, the occurrence of a great hurricane one year does not reduce the likelihood that a similar storm will strike again the next year.

In contrast, death tolls from modern hurricanes have been greatly reduced, thanks to the Weather Service warnings, radio and television communications, and evacuation plans. Nevertheless, we must not grow complacent; storm response can be improved. The hurricane watchers of the National Oceanic and Atmospheric Administration (NOAA) track hurricanes and provide advance warning for the evacuation of threatened coastal areas. Yet as

little as 9 to 12 hours of advance warning may be all that is possible, given the unpredictable turns a hurricane can take. Individuals need to be prepared, to know their community's storm-response plan, and to take appropriate action when the warning comes. Unsafe development and allowing population growth to exceed the capacity for safe evacuation must be prevented. A hurricane approaching the south Florida coast will trigger the evacuation of tens of thousands of residents and visitors from the Florida Keys into the Miami metropolitan area. These people, plus the metropolitan area population, will then need to be evacuated or sheltered. Add to this the large number of retired, elderly, and special-needs people living in the area, and the emergency preparedness and response teams will certainly be taxed to the limit. The situation is similar for the New Jersey shore, parts of the Carolinas, New England's urban corridor, and the metropolitan New York City area, which sits in the center of the New York Bight, a funnel-shaped stretch of shoreline perfect for augmenting storm surge to a frightening maximum!

The National Weather Service has adopted the Saffir/Simpson Scale (table 3.2) for communicating the strength of a hurricane to public safety officials of communities in the storm's potential path. The scale ranks a storm on three variables: wind velocity, storm surge, and barometric pressure. Although hurricane paths are still unpredictable, the scale communicates quickly the nature of the storm—what to expect in terms of wind, waves, and flooding. The risk assessment procedure and property damage mitigation recommendations presented in this book are based on physical processes associated with a moderate category 3 hurricane. Category 4 and 5 storms will cause massive property damage or destruction in spite of mitigation efforts.

Do not be misled by such scales, however. A hurricane is a hurricane. The scale simply defines how bad is bad. When the word comes to evacuate, *do it*. Wind velocity may change, or

the configuration of the coast may amplify storm-surge level, so the category rank can change. Don't gamble with your life or the lives of others.

Wind: The Universal Agent of Destruction

The strongest winds of a hurricane may exceed 200 miles per hour (320 kilometers per hour), but the maximum winds of the largest storms to hit coastal areas are rarely recorded because wind-measuring instruments are destroyed or blown away! The areas impacted by the hurricane-force winds are usually within 20 to 30 miles (30 to 50 kilometers) of the track of the eye. Frederic's winds reached 160 miles per hour (250 kilometers per hour) at sea and blew at 145 miles per hour (230 kilometers per hour) in the Dauphin Island, Alabama, area; Camille came ashore as one of the most intense hurricanes ever with devastating 190 mile-per-hour (300 kilometer-per-hour) winds. Andrew was a compact, but very intense, storm with peak winds of 180 miles per hour (290 kilometers per hour) over southern Florida.

Considering that the diameter of a hurricane ranges from 60 to 1,000 miles (100 to 1,600 kilometers) and that gale-force winds may extend over most of this area, the total energy released over the thousands of square miles covered by the storm is almost beyond comprehension. No ship or seawall, cottage or condominium, or other static structure will be immune from the impact of such forces!

Waves: Interfacing with Air, Sea, and Land

Some magnitude of hurricane, tropical storm, or northeaster impacts much of the U.S. coast almost every year. Major hurricanes that score "direct hits" are important as sediment movers, but almost any storm may have an impact and may bring torrential rain, causing river flooding as well as storm-surge flooding. Depending on the track of a given storm, waves can approach the coast from almost any direction. If the storm passes along the coast or at a low angle to the coast, the time-averaged effect means that each point along the coast will be subjected to the same forces.

The maximum height to which wind-generated waves will grow is controlled by three factors: velocity of the wind, the areal expanse over which the wind blows (called *fetch*), and the length of time the wind blows. The wave height limit is essentially the extent to which energy from the wind can be transferred by friction to the water surface. In the relatively simple case of a major winter storm, the long fetch (hundreds to over a thousand miles—up to over 1,500 kilometers) and duration (up to several days) means that the maximum wave heights for that given wind speed will develop. This condition is called a *fully developed sea.*

Estimating waves under hurricane conditions is much more difficult than for northeasters because hurricane winds vary with time and are circular and because the hurricane quickly moves over waves generated at various angles to the path of the storm (Bretschneider, 1966). In addition, fetch lengths of constant wind speed and direction in hurricanes are so small that a fully developed sea is not achieved (USACOE, 1984). The circulation of the wind and the forward motion of the storm influences the wave field development, increasing relative wind speeds in the right quadrants of the storm (relative to the track of the storm) and decreasing relative wind speeds in the left quadrants. The effect of forward motion of the storm on the wind field decreases with distance from the zone of highest winds. The maximum waves in hurricanes are generated to the right of the eye at exactly the radius of maximum winds.

Storm Surge: Landfall and Coming Ashore

Hurricanes produce storm surges, which result from the interaction of several forces. The water surface is actually set into circular motion by the fierce counterclockwise winds of the storm. This motion actually pushes water inward toward the eye of the storm and generates a convergence of water mass in the surface layer. In addition, the low atmospheric pressure in the hurricane's eye means that air pressure is not pushing down on the ocean surface at the hurricane's center with the same weight or force as on the storm's periphery. This intense low pressure causes a local rise of sea level due solely to air pressure differences. A

rule of thumb is that a 1-millibar change in air pressure translates into a change in sea level of about 1 centimeter. Not very much, but when you consider that a category 5 hurricane can have a central pressure of less than 920 millibars compared to about 1,013 millibars for normal atmospheric pressure, that means that almost one full meter of the storm surge can be due to atmospheric pressure differences as the eye crosses the shoreline!

The combination of these effects creates a mound of water, the crest of which lies to the right of the hurricane center near the position of maximum winds. The counterclockwise wind circulation within the storm combined with the forward motion of the storm itself causes the greatest energy to be concentrated in the right front quadrant (for Northern Hemisphere hurricanes). Thus, surge heights and storm effects are usually greatest to the right of the eye.

Storm characteristics are not the only controlling factors of storm-surge elevation. The nature of the offshore continental shelf has an important influence. Storm surge is augmented by shelf width, just as tidal amplitude is controlled in large measure by shelf width. On narrow shelves there simply is not enough "space" to pile up a large volume of the storm-surge water, and the water ends up "leaking out" back to sea without piling up to appreciable heights. Narrow shelves, such as those off Caribbean islands or Cape Hatteras, North Carolina, or Miami Beach, Florida, have inher-

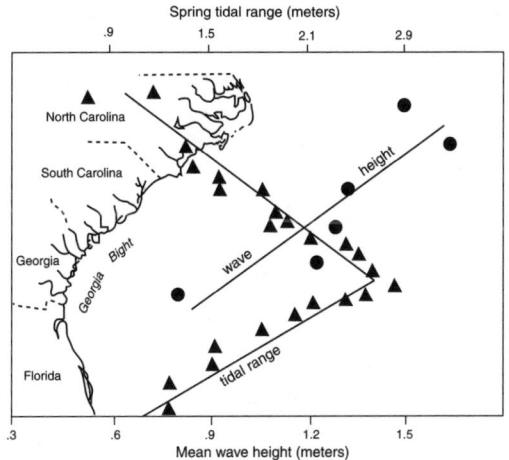

3.8 Regional variation in average wave height and tide range along the southeastern U.S. coast between Capes Hatteras and Canaveral. The high tidal amplitude and low wave height in Georgia both occur because the continental shelf is widest there. Modified from Hayes (1979).

ently lower maximum potential surge elevations. But when storm winds push surge water against a land mass, surge heights can grow to frightening proportions. On wide shelves, a broad expanse of shallow water is mobilized and piled up as storm surge, much as tides mobilize and pile up water. Storm surge piles up against the land, forcing its absolute height to elevations well above the high-tide mark, resulting in flooding. Wide shelves, such as off St. Simons, Georgia, Galveston, Texas, and St. Petersburg, Florida, have much greater maximum potential storm surges than narrow shelves, such as off Cape Hatteras, North Carolina, and Cape Canaveral, Florida. In 1969 storm surges reached 20 feet (6 meters) and possibly even 30 feet (9 meters) during Hurricane Camille on the Mississippi coast, where the shelf is very wide. The 1993 Storm of the Century blew for several days straight onshore in the panhandle area of Florida, creating storm surges of nearly 30 feet (9 meters) on the wide, very gently sloping shelf setting.

Conversely, broad shelves typically have lower wave energy because the wide shelf allows more frictional damping of wave energy as the waves traverse the shelf. Research by Miles Hayes and his students at the University of South Carolina documented the relationship between shelf width and maximum tidal amplitude and average wave height for the southeastern United States (fig. 3.8). Nevertheless, all areas subject to storm surge, including those with wide shelves, can suffer severe storm wave damage.

FIRMs and SLOSHs: Defining the Storm-Surge Risk

Storm surge is commonly measured relative to the so-called 100-year flood. The Federal Emergency Management Agency (FEMA) publishes maps showing various flood zones for the entire United States in order to identify flood hazard areas and to develop flood insurance rate maps (FIRMS) for those areas. Coastal

high-hazard areas, according to the FIRMS, are A zones (100-year-flood zone; a 1 percent chance of flood reaching or exceeding a predetermined level in any given year) and V zones (100-year-flood zone also subject to storm-driven waves). Other zones shown on FIRMS are B zones (100- to 500-year-flood zone) and C zones (>500-year-flood zone). Criteria used to define V zones and A zones are shown in figure 3.9 and are the preliminary basis for defining extreme- to high-risk zones.

A computer simulation model developed for the National Weather Service, called SLOSH (for *sea, lake, and overland surges from hurricanes*), is used to predict the still-water superelevation (that is, the storm surge) of storm waters caused by the drop in barometric

3.9 Criteria used to define FEMA A and V zones. The A and V zones are both 100-year flood zones. V zones are exposed to wind and susceptible to waves 3 feet or greater in height.

pressure, wind speed, forward speed of the storm, storm track, nearshore bathymetry, shoreline configuration, and nearshore topography. The SLOSH model is essential for developing hurricane evacuation plans in exposed coastal areas.

SLOSH models give good examples of the differences in maximum potential storm surge on narrow shelves versus broad shelves. For example, on the northern shelf of Puerto Rico, a narrow shelf, less than a mile wide, the predicted maximum storm surge from a category 4 hurricane is highly variable, ranging anywhere from 4 to 11 feet (1.2 to 3.4 meters) (Mercado, 1995). Hurricane Hugo was a category 4 when it hit Puerto Rico in 1989 and caused surges of about 3.5 feet (about 1 meter) in San Juan. Surges were less than the predicted maximum because Hugo passed over just the northeastern corner of Puerto Rico, not a direct hit that would cause

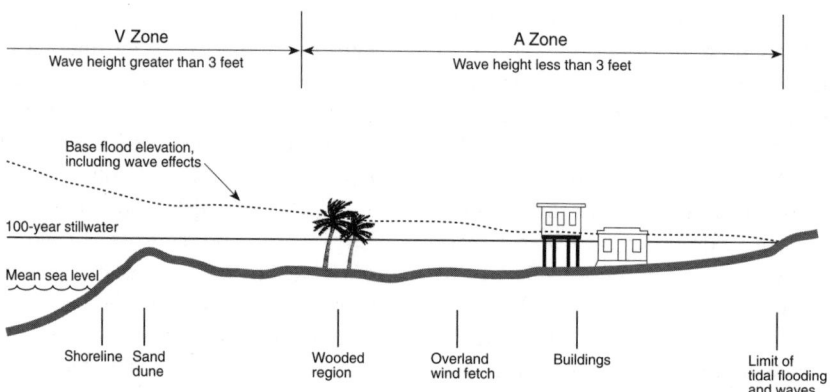

maximum surge levels (San Juan was to the left of the eye).

In contrast, Hugo hit South Carolina head-on and at high speed, both conditions contributing to maximum storm surge. Predicted storm surge from a category 4 hurricane in South Carolina is about 13 to 18 feet (4 to 5.5 meters). Measured surge elevations ranged from almost 20 feet (6 meters) in

3.10 Map showing wind circulation patterns of Hurricane Hugo at the time of landfall near Charleston, SC, on September 22, 1989. This was a coast-perpendicular storm, with onshore winds on the right side of the eye pushing waters inland and offshore winds to the left side of the eye increasing the effects of storm-surge ebb (from Coch and Wolff, 1991).

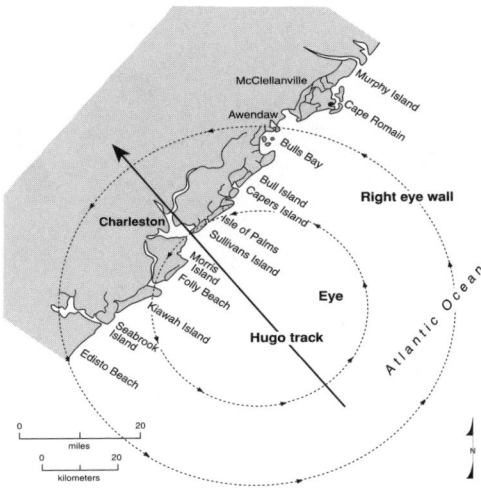

McClellandville, in the Cape Romaine area north of the eye, to 12 feet (3.6 meters) at Folly Island, south of the eye. The SLOSH models predict some of the highest potential storm surges, almost 36 feet (11 meters), for Wakulla County in northwest Florida. Together, SLOSH and FIRM define minimum elevations to be above the level of wave and flood damage, but do *not* address other hazards (e.g., wind, erosion, inlet potential).

Storm-Surge Ebb: Water Returns to the Sea

The counterclockwise circulation in tropical cyclones is responsible for abrupt changes in wind direction and intensity during storm passage. As a storm passes landward across a barrier coastline this circulation pattern produces abnormally high water-level differences between ocean and lagoon. For a hurricane moving perpendicular to the Atlantic coast toward a barrier coastline (fig. 3.10), the initial winds of the storm are onshore to alongshore. Elevated water levels occur along the front of the barrier and on the mainland side of the lagoon. After passage of the center of the storm landward, an offshore wind will result to the left of the eye and an alongshore wind to the right of the eye. If the forward motion of the storm is rapid, reversal of wind direction is abrupt, giving rise to abnormally high storm-surge ebb against the back side of the barrier island at

the same time sea level on the ocean side is low as winds push the water seaward. These conditions lead to flood flow across the island in the seaward direction, resulting in erosive scour (see fig. 3.6) and even the formation of new inlets, eroded through the island from the back side (fig. 3.11).

Such a surge can contain a large volume of water if the lagoon is large, such as Pamlico Sound or Chesapeake Bay. In a coast-parallel hurricane, as the storm arrives at the coastline (fig. 3.12a), onshore winds and storm surge are dominant. As the hurricane moves away from the area (fig. 3.12b), offshore winds and storm-surge ebb are the dominant physical processes. This is exactly what occurred during Hurricane Emily as it passed offshore of Cape Hatteras on August 31, 1993. Emily was a weak category 3 hurricane and stayed completely offshore. Winds blowing over the estuary (Pamlico Sound) caused maximum storm surge on the back side of Hatteras Island (Bush et al., 1995). Surge elevations were amplified by the concavity of the soundside shoreline of the Cape Hatteras cuspate foreland. Soundside water levels and wave heights were greater in magnitude than those at the open-ocean shoreline. The sound side of the island suffered direct wave damage, as did the ocean side. Along this entire stretch, storm-surge flooding from the sound carried mud, salt marsh grass, and other debris onto land. The rule is that back sides and interiors of islands need the same attention to planning

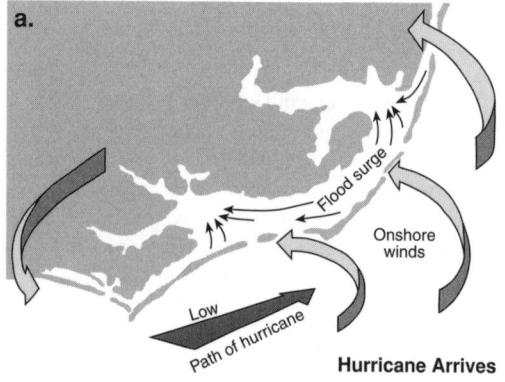

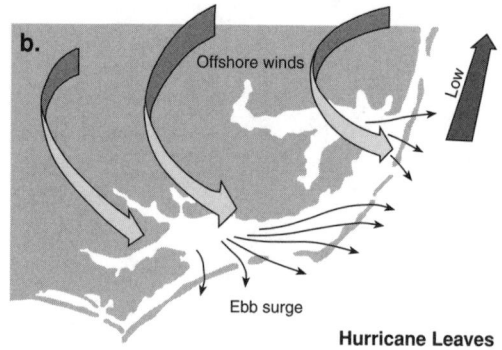

3.12 Map showing wind and surge movements of a coast-parallel hurricane along the Outer Banks of NC. (above) Hurricane arrives and the storm surge is the dominant process. (below) The hurricane moves past area, the winds reverse, and storm-surge ebb becomes dominant (after Fisher, 1962).

3.11 Inlet eroded through the Pawleys Island, SC, spit during Hurricane Hugo. Return flood flow from the channel in back of the island eroded the inlet. Photo by Rob Thieler.

and property damage mitigation as the front sides of islands. A mighty fortress (e.g., seawall) is worthless if the attack comes from the rear.

Hurricane Hugo: What We Learned

Hurricane Hugo provided an excellent opportunity to document the effects of storm-related processes on developed shorelines. Field observations were made immediately after the storm, and aerial photographs and video were taken within 36 hours after landfall. These numerous subsequent studies show that the distribution of hurricane damage was often determined by the type of development (Gayes, 1991; Lennon, 1991; Thieler and Young, 1991). Damage, for example, was particularly severe where dunes were absent, where dunes were low and narrow, where elevations were low, where vegetation was absent, and along shore-perpendicular streets and beach access sites. The latter, together with gaps between large, solid, ground-level buildings, concentrated flow, adding to the erosive power of currents. Houses that detached from foundations became battering rams, and debris missiled off buildings by the wind became additional hazards.

Some of the same lessons were learned from observations of the impact of Hurricane Gilbert (1988) on the Yucatán Peninsula of Mexico (Thieler and Bush, 1991) and of Hurricane Hugo in Puerto Rico. Likewise, a look back at previous storms and evaluation of storms impacting areas in a variety of geologic and climatic settings gives insight into shoreline response (see chapter 9).

Storm History

Hurricane Hugo was detected on satellite imagery on September 9, 1989, when a cluster of thunderstorms moved off the coast of Africa (Golden, 1990; Brennan, 1991). Hugo moved westward at 20 miles per hour (32 kilometers per hour) across the tropical Atlantic Ocean and became a tropical storm on September 11 and a hurricane on the 13th (about 1,100 nautical miles—almost 3,000 kilometers—east of the Lesser Antilles islands). Hugo passed over the Lesser Antilles causing great havoc and brushed the northeastern corner of Puerto Rico as a category 4 hurricane. It weakened somewhat upon encountering the Puerto Rico landmass but reorganized as it passed back over the Atlantic Ocean on its way north.

Hugo increased in strength over the Gulf Stream, and its forward speed accelerated.

Hugo was a category 4 storm (see table 3.2) when it made landfall in South Carolina just after midnight on September 22, 1989. Highest wind gusts occurred just before landfall and were measured at 140 miles per hour (220 kilometers per hour). The hurricane's central pressure of 934 millibars made Hugo the tenth most intense storm since 1900, when good barometric pressure records began.

Final landfall was on the South Carolina coast near Charleston at the Isle of Palms, with the eye of the storm moving northwestward at nearly 30 miles per hour (48 kilometers per hour). Moving inland and weakening, the center passed near Columbia, South Carolina. By 8:00 A.M. on September 22, Hugo had weakened to a tropical storm and passed just west of Charlotte, North Carolina. The storm moved northwest across extreme western Virginia, West Virginia, eastern Ohio, and near Erie, Pennsylvania, weakening to an extratropical storm, and tracked for two more days northeastward across eastern Canada and into the far north Atlantic Ocean.

Impacts of Hurricane Hugo

Total property damage in the coastal counties of South Carolina and southern North Carolina was estimated at near $7 billion (U.S. Dept. of Commerce, 1990). Most of the coastal damage was caused by the intense winds, waves, and storm-surge flooding associated with Hugo. Once the storm passed inland, storm-surge ebb scour left a final imprint on the coast and caused additional damage. We all hear about loss of life, houses destroyed, infrastructure damaged, and people's lives changed forever, but another impact of Hugo was the generation of tons and tons of debris—the wreckage of the storm. According to state officials, the debris generated by Hugo has shortened the lifespan of South Carolina's sanitary landfills by seven years. Seven years' worth of garbage created in just a few hours!

The South Carolina coast is located near the center of the Georgia Bight, the long concave coastal reach between Cape Hatteras, North Carolina, and Cape Canaveral, Florida. The continental shelf here is about 60 miles (100 kilometers) wide. This setting creates a situation where water can be funneled somewhat, focusing and intensifying the storm surge. Records of high-water stains and debris lines in conjunction with tidal records obtained from NOAA (NOAA/NOS, 1990) were used to ascertain the magnitude and time history of storm-surge and storm-surge ebb.

The maximum storm-surge elevations (fig. 3.13) recorded at stations closest to the hurricane track exceeded historical highest water elevations (NOAA/NOS, 1990). Note also that the greatest storm-surge height was to the right of the eye. A major contributing factor to the severe storm surge and storm-surge ebb along the South Carolina shoreline was the

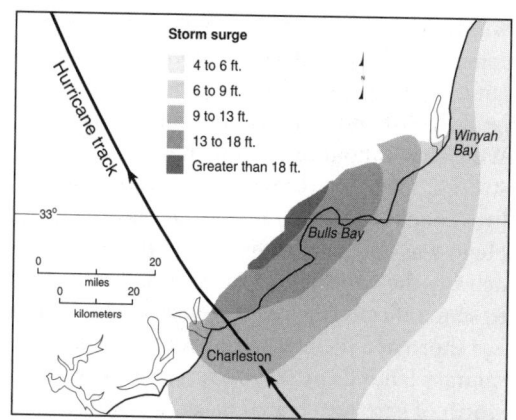

3.13 Hurricane Hugo storm-surge elevations along the South Carolina coast at the moment of crossing the shoreline (modified from Schuck-Kolben, 1990, and Coch and Wolff, 1991).

coincidence of landfall with the occurrence of high tide, causing the storm surge to be superimposed on approximately 4 feet (1.2 meters) of astronomical tide. The maximum water elevation recorded at Charleston, South Carolina, during Hurricane Hugo was 13 feet (3.9 meters). The station at Winyah Bay recorded a maximum elevation of 9.5 feet (2.9 meters). At Fort Pulaski, Georgia, located near the South Carolina–Georgia border, the maximum storm surge was 8 feet (2.4 meters). Personal inspection of high-water stains indicated a storm-surge level of 12 feet (3.6 meters) at Folly Island and Sullivans Island. Wave impact and

flood damage was extensive from Kiawah Island, South Carolina, to Cape Fear, North Carolina.

Hundreds of houses were damaged or destroyed on the islands of the Charleston area. Houses on Pawleys Island were floated off their foundations and deposited in the marsh, intact, up to a mile away. Mobile homes were gathered like toothpicks in Garden City. Water ran through the first floor of houses elevated on stilts on Sullivans Island, Isle of Palms, and Folly Island. Storm-surge flooding and waves wiped out 17 miles (28 kilometers) of dunes along developed portions of the shore. Essentially all of the approximately 8 miles (13 kilometers) of shore-protective structures were overtopped or damaged (Theiler and Young, 1991).

Hugo provided the hard lessons of experience. Frontal dunes and interior dunes were severely impacted; however, they provided valuable protection against initial storm surge, albeit temporary, reduced penetration of overwash, and protected against the damaging effects of storm-surge ebb scour. Theiler and Young (1991) concluded that the minimum dune field that survived Hurricane Hugo and thus protected buildings was about 100 feet (30 meters) wide with crests of about 10 feet (3 meters) in height. Most of the buildings damaged or destroyed by Hurricane Hugo were fronted by beaches less than 10 feet (3 meters) wide and dune fields less than 50 feet (15 meters) wide. Low dunes that were

greater than 15 meters wide were overtopped quickly and not eroded severely. These latter dunes did not prevent flooding, but reduced the impact of storm surge and waves. All shore protection structures were overtopped; none prevented flooding, and none offered any wave-breaking protection to structures located behind them! Many of these stabilizing structures (seawalls, bulkheads, groins) were damaged or destroyed, adding to the total cost of storm damage. Chapter 5 discusses the storm's implications in more detail.

Readings from six surviving tide-gauge stations indicate that storm-surge ebb was complete in less than 12 hours. The storm surge and storm-surge ebb history during Hugo is shown on figure 3.14 for Charleston, South Carolina. In Charleston, ebb occurred in two separate stages. Water level dropped 5 feet (1.5 meters) in less than one hour and then after a half-foot (0.15 meters) resurgence fell another

3.14 Time history of water elevations during Hurricane Hugo at Charleston, SC. (Adapted from NOAA/NOS, 1990).

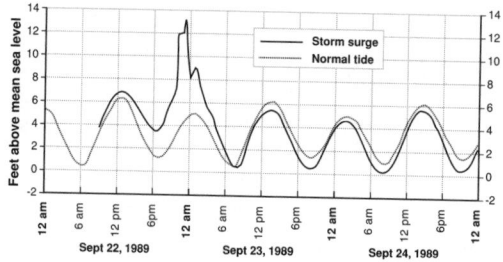

6 feet (1.8 meters) in the next 5 hours. Storm-surge ebb in this case was completed in 9 hours. The wind field associated with Hurricane Hugo is shown diagrammatically in figure 3.15. Wind is one of the primary causes of storm surge, but the perpendicular track of Hugo, its high forward speed, and the concave shape of the South Carolina shoreline were important contributing factors.

Storm-Surge Ebb Effects

Storm-surge ebb is not a dominant process in all hurricanes, but its intense erosive scouring during Hugo was responsible for property damage to buildings, seawalls, roads, and water lines. Ebb-scour channels eroded through washover deposits, indicating that the process was separate from, and took place after, the erosion and deposition induced by storm surge and waves. Storm-surge ebb scour was observed in all of the developed areas affected by Hurricane Hugo.

Storm-surge ebb scour was responsible for the creation of breaches in the narrow spit on Folly Island. One channel had dimensions of about 250 feet by over 30 feet (75 meters by 10 meters). The depth of scour channels was relatively shallow compared to scour pits. Steep-sided 6- to 12-feet-wide (about 2 to 4 meters) scour pits with depths of up to 6 feet (2 meters) and lengths of 15 to 25 feet (5 to 8 meters) cut through lawns and beach access paths. The absence or presence of vegetation

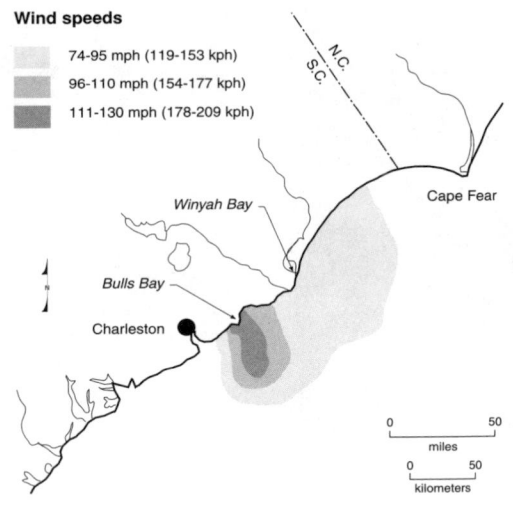

Wind speeds

- 74-95 mph (119-153 kph)
- 96-110 mph (154-177 kph)
- 111-130 mph (178-209 kph)

N.C.
S.C.

Cape Fear

Winyah Bay

Bulls Bay

Charleston

0   50
miles

0   50
kilometers

3.15 Map showing the wind categories at ground level, as Hurricane Hugo crossed the South Carolina coast. Tropical-storm force winds extend well beyond shaded areas.

controlled the size of scour pits. Once scouring was initiated, it grew to include areas where the soil and sand were not protected by shrubs, trees, or other vegetation. Again, specific development features influenced the occurrence of storm-surge ebb scour. Storm-surge ebb must be seriously considered in planned property damage mitigation, especially given that documented damage attributable to it is so prevalent: Hurricane Gilbert on the northern coast of the Yucatán Peninsula of

Mexico (Thieler, Bush, and Pilkey, 1989); Hugo in South Carolina (Gayes, 1991; Lennon, 1991; Priddy, 1991; Thieler and Bush, 1991); the Chandeleur Islands of Louisiana after Hurricane Camille in 1969 (Wright, Sway, and Coleman, 1970); and along the Alabama barrier islands after Hurricane Frederic in 1979 (Penland, Nummedal, and Schramm, 1980).

Another common effect of receding flow is the erosion of drainage channels across the beaches (fig. 3.16). Hugo's beach ebb scars were S-shaped in plan view, more strongly curved northwards, south of the hurricane eye. The northeast curvature at the seaward ends of these channels is consistent with the last direction of wind stress as the storm moved inland. Similar large linear scour channels and smaller rounded scour pits associated with Hurricane Gilbert (1988) were found landward of beaches with seawalls and revetments, suggesting a scouring effect behind the structures (fig. 3.17).

A study of Hurricane Hugo's storm-surge ebb in South Carolina found that the occurrence of storm-surge ebb scour correlated with one or more development features: (1) shore-perpendicular roads and finger canals, (2) beach access sites between multistory buildings, and (3) orientation changes in seawalls and other types of shoreline armoring (Priddy, 1991). Development enhanced the ebb-flow velocity and associated erosion.

Morphological changes on land produced

3.16 Storm-surge ebb channels on Sullivans Island three weeks after Hugo. Storm-surge ebb scour channels are located within beach access paths at terminal ends of shore-perpendicular roads. Photo courtesy of South Carolina Office of Ocean and Coastal Resource Management.

3.17 Scour channel cut by storm-surge ebb of Hurricane Gilbert (1988) on the Mexican Yucatán Penninsula. The storm-surge ebb was funneled between the walls on either side of the road in the foreground.

by the funneling of storm-surge ebb with development were extended into the nearshore. A nearshore sonar survey of the area from Myrtle Beach to Folly Island (Gayes, 1991) showed extensive shore-perpendicular scour channels on the inner shelf. These scour chan-

nels indicated strong offshore bottom currents caused by storm-surge ebb. Nearshore bars or shoals extended 160 to 250 feet (50 to 75 meters) offshore of the gaps between multistory hotels; this suggests that sand was carried offshore by storm-surge ebb currents flowing in the gaps. Extensive debris was also identified offshore of most of the heavily developed areas of the South Carolina coast. Divers identified a wide variety of debris, including sections of damaged seawalls, swimming pools, and mobile homes up to 100 meters (over 300 feet) offshore of the beach in water depths of 2 to 4 meters (about 7 to 15 feet) (Gayes, 1991). Such currents may move sediment far enough offshore to be considered permanently lost from the beach system.

As noted, the occurrence of ebb-flood ero-

sion resulted in the opening of a new inlet on Pawleys Island (see fig. 3.11). Old eyewitness accounts described the same process during earlier storms in North Carolina where portions of the barrier island coast consist of up to 40 percent of old inlet fill, suggesting that new inlet formation is to be expected (Pilkey et al., 1982). The characterisitics of the back side and interior of the island, particularly in terms of low elevation and lack of protective vegetation, are determinants in where inlets are likely to form. The expanse of open water in the lagoon or sound, river, and creek mouths behind the islands and the lack of protective salt marsh are clues to identifying points of vulnerability (see chapter 7). Finger canals and streets across low, narrow portions of an island will increase the likelihood of new inlet formation.

Where inlets migrate, the lengthening spit end of the updrift island ultimately will be cut off when the new inlet re-forms in the earlier inlet position—nature's way of maintaining more efficient drainage from the back side of the island. Artificially closing these newly formed inlets maintains a level of inefficiency in the drainage, increasing the likelihood of another round of new inlet erosion in the next big storm and another round of property loss. Those who own the property on the Pawleys Island filled inlet can expect nature's instant replay in a future hurricane. America's barrier islands are peppered with relict and historical inlet positions. Some closed naturally and stabilized as island environments evolved; others remain as topographic lows, vulnerable to future episodes of ebb-scour erosion.

Such potential inlet positions may be stabilized by augmenting the topography (e.g., constructing dunes) and planting vegetation.

Role of Vegetation

Hurricane Hugo once again demonstrated that a heavy cover of vegetation reduces overwash penetration and storm-wave damage (Bush and Pilkey, 1994; Thieler and Bush, 1991). On Pawleys Island, South Carolina, neighboring houses suffered contrasting degrees of damage, the poorly vegetated properties being heavily damaged or destroyed; less wind damage occurred to houses located behind or within the maritime forest on the same island. Similarly,

during Hurricane Gilbert (1988) in Mexico the devegetated areas were heavily overwashed and buildings damaged (fig. 3.18). As noted, poorly vegetated areas are more susceptible to ebb scouring.

Extratropical Storms: Don't Write Them Off

A variety of nonhurricane or extratropical storms affect all U.S. coasts. These winter storms are called northeasters, southwesters, and other names depending on the direction the wind blows and on local customs. The big storms are typically given unofficial names, usually related to the date of their arrival. Two of the biggest storms in the U.S. East Coast's modern history were named the Ash Wednesday storm of 1962 and the Halloween storm of 1991. A large storm during December 1992 didn't fall on any day of particular note and was left with the designation the No-Name Storm. The northeaster/southwester of March 1993 was called the Storm of the Century, and although it was not a record-breaking coastal storm, the damage was widespread: storm surge and wave erosion in the Florida panhandle, record-breaking snow cover inland along the East Coast, and cosmetic but costly wind damage to coastal buildings. Communities such as Sunset Beach and Topsail Beach, North Carolina, were littered with the shingles from stripped roofs, and property owners scrambled to get tarpaulins over the bare roofs

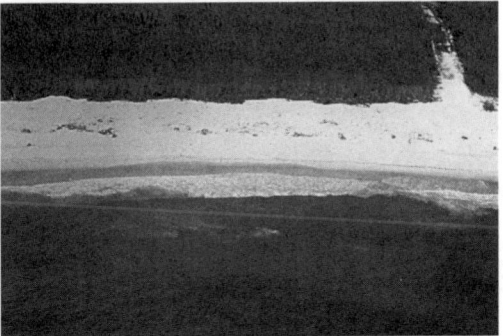

3.18 Overwash penetration is greatly increased where vegetation is removed for development. In these two Hurricane Gilbert examples from the Yucatán Penninsula, Mexico, clearing forest for a road (a) and for a condo complex (b) has allowed overwash to penetrate beyond the frontal dune.

3.19 Photo of waves from the 1991 Halloween storm engulfing the lighthouse at the entrance to Nassau Harbor, Bahamas. By some measures, the Halloween storm may have been the largest Atlantic northeaster in this century, although the 1962 Ash Wednesday storm caused far more property damage.

3.20 Overwash sand across a road at Isabela, Puerto Rico, on the north coast. The storm event was in mid-December 1991.

to prevent water damage from rainfall. The storm also tore out fiberglass insulation from underneath elevated homes and covered some sections of islands with a snow-looking blanket of fluff.

Anatomy of a Winter Storm

Hurricanes get the headlines, but for sheer size, number, and duration, winter storms are the major threat to coastal property and the major cause of shoreline erosion. From the Gulf of Alaska and along the Pacific's shores through the Gulf of Mexico (southwesters) and all along the Atlantic seaboard (northeasters), winter storms cause widespread property loss almost every year. Much of what we know about these major storms comes from the studies of Robert Dolan and Robert Davis of the University of Virginia (e.g., Dolan and Davis, 1992; Davis and Dolan, 1993; and Watson, 1993).

Extratropical cyclones, referred to here as northeasters, as distinguished from tropical cyclones (or hurricanes), form in midlatitudes.

The ideal breeding grounds for northeasters are coastal areas where there is a large temperature gradient in the wintertime between air over the cold continental land and the relatively warmer ocean waters. The low-pressure systems (cold fronts) we see sweeping across the nation every day on the evening weather can intensify as the center moves over the Atlantic Ocean.

The key to a low-pressure center growing into a full-fledged northeaster is the front (that is, temperature gradient) between polar air from the north and tropical air from the south. As a northeaster tracks up the East Coast or across the Gulf of Mexico, the storm intensifies by picking up energy from the relatively warmer waters. The damaging winds blow from the southwest in the Gulf of Mexico or the northeast in the Atlantic and are usually slowed somewhat by the forward motion of the storm center in the opposite direction. (Remember that winds are named for the direction *from* which they blow, so a northeast wind blows from the northeast, but a northeast-moving storm moves *toward* the northeast.) The presence of a strong stable high-pressure center in eastern Canada will block a northeaster from moving very quickly up the East Coast, providing the storm more time to gather strength from the ocean waters.

Northeasters typically cover larger areas, have much lower wind speeds, and move slower than hurricanes, sometimes remaining off a coast for several days. Unlike fast-moving hurricanes, the winds and waves of a northeaster may persist through several tidal cycles, amplifying the shoreline damage due to waves at any given location. Like California waves generated far away in storms off the Gulf of Alaska, the waves of an East Coast storm may travel far from the storm center and arrive at the beach on bright sunny days with only light local sea breezes in evidence (figs. 3.19 and 3.20). Generally, northeaster winds are not the major cause of property damage, and loss of life is usually low. A major exception to this was the great 1953 storm that struck Holland, breaking many dikes and leaving more than 1,500 people dead in its wake. More important, as a rule, is the direct impact of waves on buildings, the extensive flooding, and the shoreline erosion undercutting building foundations.

Table 3.3   The Dolan/Davis Northeaster Intensity Scale

Storm Class	1 (Weak)	2 (Moderate)	3 (Significant)	4 (Severe)	5 (Extreme)
Beach Erosion	Minor changes	Modest: confined to lower beach	Extends across entire beach	Severe beach erosion and recession	Extreme beach erosion (up to 50 m in places)
Beach Recovery	Full and usually immediate	Full	Usually recovery over considerable time (months)	Recovery seldom total noticeable changes	Permanent and clearly
Dune Erosion	None	None	Can be significant	Severe dune erosion or destruction	Dunes destroyed over extensive areas
Dune Breaching	No	No	No	Where beach is narrow	Widespread
Overwash	No	No	On low-profile beaches	On low-profile beaches	Massive in sheets and channels
Inlet Formation	No	No	No	Occasionally	Common
Property Damage	No	Minor, local	Loss of many structures at local scale	Losses of structures at community level	Extensive regional scale: millions of dollars

Source: Davis and Dolan, 1993.

Ranking Northeasters

An intensity scale for U.S. Atlantic coast northeasters, similar to the Saffir/Simpson Scale for hurricanes, was developed by Dolan and Davis (1992; Davis and Dolan, 1993) (table 3.3). The Dolan/Davis classification is not based on wind velocity but on the size of the waves and the duration of the storm and is expressed in terms of intensity of property damage. This classification is intended for use along the U.S. Atlantic seaboard. Since 1960, eight class 5 northeasters have occurred. On

3.21 Buxton Inlet, NC, opened by the Ash Wednesday 1962 northeaster, just north of the Cape Hatteras Lighthouse. The inlet was briefly bridged before being filled in by the highway department.

day storm, a class 5 northeaster. It struck during spring tides, resulting in extreme storm surge. Because the storm persisted over five high tides, the damage grew more and more extensive. Beachfront communities from Fire Island, New York, to Nags Head, North Carolina, were devastated (see fig. 1.4). Some damage occurred to all beachfront communities between southern Massachussets and northern Florida. Damage was particularly severe to beachfront buildings. The loss of beaches was so severe that the 1962 storm marked the entry of the U.S. Army Corps of Engineers into the arena of beach replenishment. Between 1962 and 1965 the Corps pumped in millions of cubic yards of sand to widen the beaches of New Jersey, many yards of which had actually substantially recovered through natural processes. A new inlet was opened along the Outer Banks of North Carolina (fig. 3.21).

As the nation enters the twenty-first century every new hurricane and winter storm will indeed be the "storm of the century" until the next storm breaks that record. If the impact of these yet unborn storms is to be blunted, the most thorough risk assessment and mitigation programs possible must be undertaken and emplaced.

the average since the early 1980s, about two dozen northeasters have occurred each year along the U.S. Atlantic coast (Davis and Dolan, 1993).

What Happens in a "Storm of the Century"?

The most memorable storm of this century in the United States was the 1962 Ash Wednes-

Mapping the level of relative property damage potential for barrier islands assists community officials in planning for and reducing impacts from natural disasters and allows individual property owners to choose sites and purchase property in a more knowledgeable way. Because vulnerability is determined by natural processes and environmental settings, nature should be our guide in mitigation methods, rather than our relying solely on engineering and social regulation (e.g., seawalls, building codes, zoning regulations). A useful approach is to classify risk into categories such as extreme, high, moderate, low, and even very low in some cases (table 4.1). All barrier islands have a high element of risk, and any given island may not have areas that fall into all four categories. In fact, low-elevation barrier islands, such as Dewees Island, South Carolina, may fall entirely into the extreme-risk category. Table 4.2 summarizes field evidence useful to determine risk of property damage.

Virtually all coastal areas are at extreme risk if struck by a category 5 hurricane (see Saffir/Simpson Scale, table 3.2)! Differentiation into risk zones is useless for such a storm. For a category 4 or lower storm, however, risk mapping can be meaningful. The risk mapping presented here is based on the risk afforded by a moderate category 3 hurricane, hitting directly at the site under question. A moderate category 3 hurricane will have winds of about 120 miles per hour

(190 kilometers per hour). Typical accompanying storm surges will range from 5 to 12 feet (1.5 to 3.6 meters). A category 4 storm as a basis for mapping would assume higher energy conditions and require stricter limits on risk categories.

Note that this risk mapping involves *only* risk of property damage and is *not* concerned with risks to inhabitants. In general, of course, areas with high potential for property damage are also areas of high risk for human inhabitants, but a low-risk site also can be a death trap in a hurricane. Difficulty in evacuation is an example of a human risk which may be entirely independent of homesite safety. For example, drawbridges may be stuck in the open position; evacuation routes may be blocked by flooding, downed trees, wrecks, or cars that run out of gas, with many miles to go to escape the flood zone (as on the west Florida mainland).

Storm Experience Sets New Directions

Observations of several barrier island communities after Hurricanes Gilbert (1988) and Hugo (1989), as well as several smaller hurricanes and numerous winter storms, suggest that property damage potential can be lessened significantly by prudent site selection and proper location of structures on property. Storm impacts can also be greatly reduced by islandwide mitigation based on risk mapping.

Barrier island risk assessment can be done at any scale: islandwide, individual communities on one island, neighborhoods within a community, a block within a neighborhood, and for individual structures and building sites. Naturally, property owners focus on site evaluation, as do many mitigation plans. But working from an islandwide base map of risk potential down through the community level to neighborhoods and finally to individual sites will provide a better basis for developing mitigation plans.

Determining relative hazard ranking of large sections of barrier islands is fairly straightforward (table 4.3). Important determinants of damage potential for a given structure include its elevation above sea level; elevation above ground level; exposure to wind and wave hazards (presence or absence of thick maritime forest or shrub cover; presence or absence of high, wide dune fields); and distance from the ocean or sound, which determines the likelihood of impact by storm-surge waters. Using these criteria it is possible to define areas of islands with similar risk. Response to previous storms in an area also gives critical insight into risk mapping.

Coastal Environments and Associated Hazards

The concept of different parts of a barrier island or other coastal environment responding differently to the same storm is fundamental in

Table 4.1 Characteristics of Coastal Property Damage Risk Categories (based on likely changes from a moderate Category 3 hurricane)

	Categories of Evaluation							
Risk	Primary Hazard Controls on Vulnerability (determine storm surge, flood, overwash, and wind hazards)		Secondary Controls on Vulnerability (other modifying hazards)				Probable Damage/ Destruction in Category 3 Hurricane or Equivalent Wind/Wave/Surge	Mitigation Recommendations
	Elevation	Vegetation	Erosion Rate	Inlet Potential	Construction Factor	Vulnerability		
EXTREME	Low; mostly in V zone; some A zone sites. Typically back beach, low frontal dune, active overwash front, similar on soundside and near inlets	None or sparse beach/dune grasses only; no maritime forest or shrub thicket	High to moderate	Near migrating, historic, or potential inlet position	Older buildings, built before building codes, not to code, or with code violations, closely spaced, or new buildings that were not built to code or violate code	MAXIMUM (likely to be impacted by 4 or more hazardous processes)	Total destruction to very heavy damage by direct wave attack for all elements of buildings (roof, windows, walls, foundations, decks, porches, stairs, garages) as well as outbuildings, utilities, services, and landscaping	Relocate before storm or abandon site after storm. *Do not rebuild destroyed buildings in place.* Protect marshes. Preserve, augment, and restore all natural environments
HIGH	Low; in A zone and lacks forest or shrub thicket cover. Typically overwash apron or fan extension but away from V zone; inner dune trough or frontal blowout in dunes, perhaps flat interior grasslands	Sparse to none, or greatly disturbed by development or past salt-water kills	High to moderate	As above	As above	HIGH (likely to be impacted by at least 3 hazards)	Total destruction is not uncommon and heavy damage is likely; wind damage most probable; all building elements at risk (roof, windows, walls, foundations, attachments) as well as outbuildings, services, and landscaping	Relocation is most prudent; elevate and vegetate; build protection in region around site (e.g., dune projects, forestation); plug dune gaps; change street layout; maintain existing projects (e.g., beach nourishment), but not shore hardening. Protect marshes

Table 4.1 continued

	Categories of Evaluation							
Risk	Primary Hazard Controls on Vulnerability (determine storm surge, flood, overwash, and wind hazards)		Secondary Controls on Vulnerability (other modifying hazards)				Probable Damage/ Destruction in Category 3 Hurricane or Equivalent Wind/Wave/Surge	Mitigation Recommendations
	Elevation	Vegetation	Erosion Rate	Inlet Potential	Construction Factor	Vulnerability		
MOD-ERATE	Moderate to low and *not* in V zone; possibly A zones if heavily forested. B zone or out of flood zone (but down-graded by sec-ondary hazards). In dunes or other elevated land-forms such as built-up terrace	Mainland forest cover (buildings under canopy level)or shrub thicket (not seri-ously disturbed by development), interior and fron-tal dunes well-vegetated, is-land backed by healthy marsh. A zones even with good vegetative cover can still be at high risk	Low to accre-tionary	Should be away from migrating inlet; downgrade if near his-toric or potential new inlet	Downgrade if as above. All build-ings should meet or excede build-ing code stan-dards	MODERATE likely to be impacted by at least 2 hazards, and nui-sance haz-ards such as over-wash and dune mi-gration are likely)	Heavy to moderate damage should be expected. Wind damage very likely (e.g., windows, roof, walls, at-tachments). Flood damage less likely. Landscaping af-fected by overwash and blowing sand. Above ground services likely to be interrupted	Elevate and veg-etate on site sur-rounding region (e.g., vegetate dunes, rehabilitate forest, plug over-wash gaps, or work to maintain natural overwash sites; promote sand additions to island interior; modify street lay-out. Protect marshes.
LOW	High, C zone (or B zone but upgraded by healthy maritime forest and no secondary prob-lems)	Maritime forest cover is healthy and largely undis-turbed. Surround-ing environs also well-vegetated	Zero to accre-tionary	No inlet potential	All construction is at or above code and is well-maintained	LOW (no more than one hazard likely)	Expect damage, but not heavy. Wind damage is most likely (cosmetic to ex-terior roof, walls, win-dows, attachments). Poten-tial serious damage from falling trees or blowing de-bris. Rain damage if glaz-ing fails or roof/wall leaks develop	Augment protec-tive vegetation and dunes as above. Do not remove sand. Protect marshes.

Table 4.2   Field Parameters to Determine Category of Risk of Property Damage in Coastal Storms*

Geo-Indicator	High Risk	Moderate Risk	Low Risk
Site elevation	< 3 m	3 m–6 m	> 6 m
Erosion/Accretion rate	Severely to slowly eroding	Stable	Accreting
Beach width, slope, and thickness	Narrow and flat, thin with mud, peat, or stumps exposed	Wide and flat, or narrow and steep	Wide with well-developed berm
Overwash	Overwash apron (frequent overwash)	Overwash fans (occasional overwash)	No overwash
Site position relative to inlet or river mouth	Very near	Within sight	Distant
Dune configuration	No dunes (see Overwash)	Low, or discontinuous dunes	High, continuous, unbreached ridge
Bluff (unconsolidated) configuration	Bare face, recent or no talus ramp	Vegetated face and well-developed ramp	Low slope angle (large ramp), mature cover of vegetation
Coastal shape	Concave or embayed	Straight	Convex
Vegetation on site	Little or toppled vegetation	Well-established shrubs and grasses, none toppled	Mature vegetation, forested, no evidence of erosion
Drainage	Poor	Moderate	Good
Area landward of site	Lagoon, marsh, or swamp (e.g., mangrove)	Floodplain or low-elevation terrace	Upland

* Parameters are generalized and must be modified for individual islands.

developing any approach to hazard recognition and risk assessment (see tables 2.1, 3.1, 4.1, 4.2, and fig. 2.3). For example, during a storm, low-elevation environments, such as beaches and low-elevation dunes, will be subject to high wind, storm-surge flooding, and direct wave attack. During the same storm, higher elevation dunes and maritime forests may only experience heavy rainfall, associated surface runoff of the rainwater, high wind, and possibly flooding in some maritime forest areas depending on elevation (see figs. 2.3 and 2.4).

Elevation and exposure to wind (forest or dense shrub thicket cover) are the most important characteristics of the natural setting from the standpoint of potential for property damage. Stable higher elevation sites have lower potential for flooding and wave attack. Sites with lower exposure to wind because of the presence of dense maritime forest or nearby high, wide dune fields also are at significantly lower risk. High elevation may also mean greater exposure to wind forces. However, wind is much easier to engineer for in buildings than are waves or the forces of rising water. In large measure, then, the higher elevation and more densely forested portions of islands have lower potential for damage from storms and are rated as moderate- to low-risk sites, depending on secondary factors. These and similar criteria provide a basis for delineating portions of islands with similar risk or damage potential as low, moderate, high, or extreme. In general, most barrier island environments fit neatly into these categories (fig. 4.1 and table 4.4). Of course, some local variation in these categories should be expected. For example, high elevation at Nags Head, North Carolina, may be less than the lower limits of high elevation on a Texas island where storm surge would be higher. In other words, the overall classification is *qualitative*, and absolute quantitative limits for determining parameters such as elevation and percent or density of forest cover must be determined for each region, again based on storm experience and the criteria established for developing FIRMs and applying predictive models such as SLOSH. The same is true for additional "qualifier" hazards (e.g., erosion rates, inlet migration rates, potential for inlet formation, type and quality of existing development, engineering structures, dune width and height). Absolute quantification may be debatable; however, when dealing with barrier islands the qualitative approach is a very good first approximation of risk level based on past experience. Application of this risk assessment/mitigation technique is relatively simple, quick, and inexpensive, providing both individuals and community officials with a useful evaluation tool.

Location of a coastal environment relative to the open-ocean shoreline can influence risk potential and complicate the ranking scheme. For example, a low-risk maritime forest can be exposed on the beach in a high-erosion situation (fig. 4.2), leading to its classification as high risk. The general assumption underlying

Table 4.3   Coastal Risk Assessment Method

I. Primary Factors		Decreasing Vulnerability $\longrightarrow$		
Elevation (rating factor)	V zone* (5)	A zone (4)	B zone (3)	Above B (2)
Vegetation [wind exposure] (rating factor)	None or sparse shrub (3)	Dense shrub thickets (2)	Maritime forest (1)	
Preliminary rating (sum of factors)	Extreme (≥9)	High (8–7)	Moderate (6–5)	Low (4–3)

II. Secondary Factors (Add 1/2 point per factor)	(Subtract 1/2 point per factor)
+High erosion rate	−Accretion or rocky shore
+Migrating, historical, or potential inlet	−Dune field wide as well as high
+Severe historical storm impact	−Area weathered all past storms well
+Development hazards	−Development design with nature
+Engineering concerns	−Stabilizing structures
+Other	−Other

Revised risk rank (total of factors)	Extreme (≥9)	High (8–7)	Moderate (6–5)	Low (4–3)	Very low (≤2)

*Property in any V zone is considered to be at extreme risk.

Table 4.4   Typical Environmental Settings within Risk Categories of Southeastern U.S. Barrier Islands*

Low risk:
  1. Mainland forest (high elevation)
  2. Maritime forest (high elevation)

Moderate risk:
  3. Maritime shrub thicket (high elevation)
  4. Vegetated interior dunes (high elevation)
  5. Active dune fields (high elevation)
  6. Vegetated interior dunes (varying elevation)
  7. Dune swales and blowouts

High to Extreme risk:
  8. Overwash apron (low elevation, no forest)
  9. Washover fans (low elevation, no forest)
  10. Frontal dunes
  11. Ocean beaches

* The importance of these hazards may vary from island to island.

the definitions of these parameters is stability (e.g., the maritime forest isn't falling over an eroding scarp; the protective dunes are grass covered and not blowing out).

Finally, climate must be taken into account. The nature of the vegetation cover will vary from tropical (rain forest, mangroves) to temperate (maritime and mainland forest, as in this discussion) to arid (grass and shrub cover). In the last case, the wind hazard is enhanced by lack of forest cover, but if dune formation has been active, high elevations may be present.

Hazard Zones

Extreme Hazard

Areas rated as *extreme-hazard areas* are the lowest elevation portions of islands and those most exposed to wave-generated winds during storms (see table 4.1 and fig. 2.3). By definition they are within the 100-year flood level and exposed to potential wave-generating winds enough to have waves greater than 3 feet formed on top of the possible storm surge,—that is, in V zones as defined on FEMA flood insurance rate maps (FIRMs, discussed later in this chapter, are available in any city hall and from FEMA; see appendix for address). Extreme-hazard zones typically have little vegetation except sparse growths of low beach or dune grasses. There is no maritime shrub thicket or forest, either never existing or

having been removed for development. These zones experience flooding from storm-surge waters as defined by the SLOSH model and from heavy rains; are likely to be overwashed during storms; are susceptible to wave attack; usually affected by erosional shoreline retreat; and may experience storm-surge ebb scour. In addition to oceanfront areas within the FIRM V zones, island areas adjacent to inlets, lagoons, and estuaries as well as low, barren areas such as overwash terraces located away from the shoreline may fall into the extreme hazard category.

High Hazard

High-hazard zones are typically low-elevation areas within the 100-year flood plain (FIRM A zone) and typically have no or poor vegetative cover (see table 4.1 and fig. 2.3). Areas within maritime forest and/or shrub thicket may also be classified in this high-hazard zone because of susceptibility to flooding and even wave attack even though maritime forest is important for reducing exposure to high wind. The low elevation creates vulnerability to flooding from storm surge and heavy rains. Potential for wind damage remains high. These zones are usually located more toward island interiors and are usually less susceptible to erosion damage. Such areas may be susceptible to sound-side erosion as well as flooding from the back side of the island and locally to new inlet formation. If forest is present, the likelihood of impacts from both wind and storm-surge ebb scour is reduced.

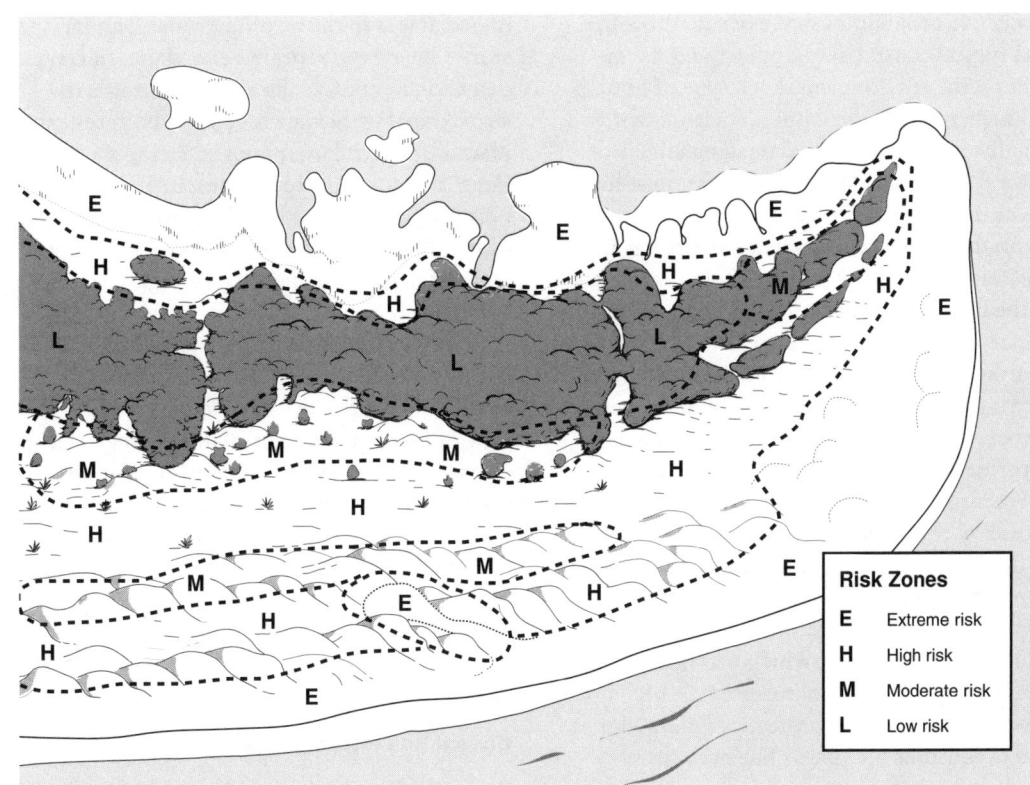

direct wave attack. However, flooding should not be ruled out (i.e., these areas may be in FIRM B or C zones, that is, having potential to be flooded in a *greater* than 100-year flood) because unusually high water levels can be generated in a category 3 hurricane. In some cases an A zone may be assigned to the moderate category if there is a good cover of vegetation, few secondary hazards, and additional natural protection like a fronting of a wide dune field. Flood insurance is recommended in

4.2 The beach at St. Catherines Island, GA, covered by fallen trees from the bluff, undermined by shoreline erosion. No buildings exist along this stretch of shoreline, hence no shoreline erosion "problem" exists. In other words, we who build buildings next to the shore create the erosion problem!

Risk Zones	
E	Extreme risk
H	High risk
M	Moderate risk
L	Low risk

4.1 Risk (for property damage) classification for an undeveloped Pandora's Island based on elevation, vegetation, and secondary factors described in the text. Risk classification changes as development proceeds. For example, the presence of a finger canal and road (fig. 3.1) led to new inlet formation in the hypothetical storm response of Pandora's Island.

Moderate Hazard

Moderate-hazard zones generally include areas above the 100-year flood zone, although often without maritime forest or dense shrub thicket cover (see table 4.1 and fig. 2.3). The high elevation means these areas are not generally subject to flooding and are unlikely to suffer moderate-hazard zones. Wind damage is the most likely hazard; the lack of a protective forest means a high degree of exposure to wind as well as to missiling.

Low Hazard

Low-hazard zones are well-forested areas that are above the 100-year flood zone (see table 4.1 and fig. 2.3). These vegetated areas are generally not subject to flooding or wind hazards. Typically these areas are inland, away from wave erosion and potential overwash or inlet effects. Sites for development within this zone are "least risky" in terms of potential for property damage. A caveat is that removing forest for development obviously reduces the amount of protection and often leads to increased degradation of newly exposed portions of the forest from salt spray. Another caveat is that building in the forest could result in damage from falling trees and limbs during storms, but most maritime forests are relatively resistant to wind destruction. Furthermore, a few holes in the roof are better than a destroyed home. As with the previous category, the rare flood event cannot be ruled out, especially if a long-term mitigation view is taken.

No location on a barrier island or coastal area can be called truly safe for development. Moreover, the potential for property damage is often increased by human alterations to the coastal environment that reduce the natural

protective capabilities that existed. Table 4.5 lists typical coastal development and its impact on the environment. Each type of human adjustment alters the natural environment, usually exacerbating the damaging effects of natural processes. Mitigation plans must look beyond site-specific requirements or island-front approaches and consider all aspects of the island and island chain. Mitigation efforts in the interior of a barrier island may be as important as on the shoreline. In the final analysis, if low risk to life and limb and property is a high priority for a potential property owner, a carefully selected mainland site is a superior choice.

Potential for property damage is high in almost any coastal setting and will only increase as development density increases or as single-family homes give way to multifamily dwellings and commercial establishments. Consider Gulf Shores, Alabama, which was heavily damaged by Hurricane Frederic (1979). Frederic eroded away protective frontal dunes and oceanfront buildings. The oceanfront single-story buildings, once protected by dunes, were replaced by larger multistory buildings in the extreme- and high-risk zones. These buildings suffered moderate to severe damage in Hurricane Opal in 1995. If a new dune line had been allowed to form, its location would have been inland where the new buildings were located! The final result is greatly increased development density, population density, and the potential for greater

financial loss in the next hurricane. Similar stories have been repeated elsewhere. In Garden City, South Carolina, small cottages destroyed by Hurricane Hugo have been replaced by much larger houses, facing a duneless, storm-narrowed beach. This development pattern is taking place over the entirety of barrier islands, often proceeding into areas of higher and higher risk (e.g., inlet zones).

The rule is that coastal risk assessment is not a one-time exercise. Risk mapping is an evaluation for the specific conditions at the time of mapping. Barrier islands change at a rate much faster than mainland environments, both in terms of natural alteration and economic development. Therefore, risk mapping must be repeated periodically, particularly after big storms, and the mitigation plan altered accordingly.

Coastal Risk Mapping Technique

Responsible officials and planners, conscientious developers, and individual buyers and property owners face a multitude of processes on barrier islands, all of which carry varying levels of risk to property. These processes, and their resulting landforms, become risk assessment parameters (tables 4.6 and 4.7). But which ones are most important? Which of these parameters are the best measure of the level of risk potential? And at what scale and

Table 4.5   Impact of Development on Natural Environment and on Risk of Property Damage

	Development Type	Direct Effects	Indirect Effects

Building site modification:	grading
	paving
	dune removal
	paths through dunes
	forest removal
	roads and other
	  infrastructure
		Changes landform configuration
		Eliminates sources of sediment
		Decreases sediment stability, changing
		  rates of on/off-island sediment exchange
			Establishes property in hazardous areas
			Increases exposure of property to wind
			  and wave hazards

Construction of buildings:	single-family
	multifamily
	high-, medium-, low-
	  rise multifamily
		Alters wind patterns
		Obstruction to sediment flow
		Truncates beach or dune zone
		Channels storm surge and ebb flow
		Reflects waves
			Focuses human use and human impacts
			Leads to construction of support
			  infrastructure and implementation of
			  protection structures
			Increases population density

Hard shoreline stabilization:	seawalls
	groins
	jetties
	breakwaters
		Armors shoreline
		Alters sediment flows
		Changes location of erosion/deposition
		  and its severity
		Reduces or prohibits on/off-island
		  sediment exchange
			Leads to further development, putting
			  more property at risk
			Leads to need for more structures
			Destroys recreational beach

Soft shoreline stabilization:	beach nourishment
	dune building
	sand fencing
	bulldozing
		Changes sedimentation rates and
		  severity of erosion
		Interferes with on/off-island
		  sediment exchange
			Masks nature of erosion problem
			Leads to further development

Inlet channel alteration:	dredging
	relocation
	artificial closure
		Alters natural current pattern
		Changes location of erosion/deposition
		  and its severity
			Can encourage development in inlet
			  hazard zones
			Leads to calls for structural improvements

Source: Adapted from Nordstrom, 1987.

Table 4.6 Secondary Factors in Property Damage Risk Categorization

Natural Risk Factors

Island Setting
1. High erosion rate increases damage potential
2. Dunes narrower than 100 feet (30 meters) and lower than 15 feet (5 meters) increase damage potential
3. Very narrow island increases damage potential
4. Historical inlet areas increase damage potential
5. Very wide lagoon increases damage potential
6. Steep slopes reduce inland penetration of surge
7. Rocky shoreline may decrease damage potential

Historical Storm Response
1. Extensive overwash penetration and areal extent increases likelihood that it will recur
2. Changes in inlet size and shape affecting large areas increase likelihood they will recur
3. New inlets opened in the past means the likelihood for recurrence is high

Human-Induced Risk Factors

Development Patterns
1. Surrounding homes built at ground level increase damage potential
2. Surrounding homes of unreinforced cement block or brick construction increase damage potential
3. Roads built perpendicular to the shoreline

Human-Induced Risk Factors (continued)

and/or cutting through dunes increase damage potential
4. Paths or road gaps found between or through sand dunes/vegetation increase damage potential
5. Large buildings that provide "wave shadow" may decrease storm damage
6. Mobile home parks increase damage potential
7. Older developments (pre-flood insurance regulations) may be less floodproof: poorly constructed; ramrodding likely; closed-in ground floors provide debris for damaging nearby structures

Shoreline Engineering
1. Large seawalls or revetment reduces damage potential to property but results in beach degradation and will cause erosion of neighboring properties down the beach
2. Breakwaters may reduce damage potential to property but will cause erosion of neighboring properties down the beach
3. Proximity to updrift jetties which may cause erosion will increase damage potential
4. Groin fields may indicate long-term erosion problem and an attempt to trap sand along one stretch of the shoreline; while so doing, they starve the downdrift portion of the shoreline of sand

level of detail should such risk analysis be carried out? To take into account all of the parameters in table 4.7 would be a daunting task, and many of the factors listed are difficult to quantify.

Our objective is to reduce the number of critical parameters to a few broadly meaningful measures of risk potential, produce a preliminary risk map, and then refine the risk evaluation based on added levels of related factors that are of more local extent and impact. Given that the major processes are wind, waves, storm surge, and flooding, two universal controls emerge as most important: elevation and protective vegetation cover (primarily forest). How the elevation is distributed in terms of landforms (e.g., dunes, overwash, interior beach/dune ridges, troughs) can be evaluated for local refinement, say an area within a barrier island. Parameters that are more site specific (e.g., erosion rates, potential overwash passes, inlet position) are incorporated at the neighborhood or individual property level for final risk evaluation.

This simple-to-complex approach allows property risk mapping to be carried out at various scales, beginning with the overview level of the entire island before trying to be site specific. As noted, the approach is qualitative, which allows application with a minimum of background, is cost efficient, and can be communicated to the most important audience, the citizens of the island in question. At the same time, quantitative limits on many of the parameters can be assigned for individual islands or island chains, based on past storm experience, and ultimately used to define mitigation regulations. Some of these quantifiers are already in use in the form of FIRMs (the 100-year flood level plus a calculated wave height is the regulatory elevation) and storm-surge flood elevations based on the SLOSH model. Minimum dune heights and widths to sustain storms and protect property are known for specific storm conditions (e.g., Hurricane Hugo in South Carolina or Hurricane Frederic in Alabama). Specific erosion rates may be designated and utilized as in calculating setback requirements. Of course, these "quantifiers" must be measured and monitored frequently to be applied usefully, adding the requirements of expertise, time, and cost. Other parameters can be quantified simply as present or absent (e.g., overwash gaps, vegetation, inlets), and some factors are difficult to quantify (e.g., what is the most effective measure of forest cover?). Ultimately, the broad list of parameters (see table 4.7) must be addressed if the overall island mitigation activities are to be successful.

The following sample mapping outline provides a model that can be adopted for a varying set of assumptions or objectives. The approach could be upgraded to a lower level of acceptable loss, for example, by defining the level of risk for a category 2 hurricane or a 200-year flood level. Similarly, downgrading might allow a higher level of acceptable loss: adopt a 50-year flood level or the conditions

of a category 4 hurricane. Other factors on the list could be given more weight where they are deemed more important (e.g., overwash and inlet potential for narrow islands). The community of Nags Head, North Carolina, is used to illustrate the map-making and risk assessment procedure. Table 4.3 outlines the procedure.

Preliminary Analysis Map

Keep in mind that the following mapping is based on island response to a moderate category 3 hurricane (see table 3.2). The most important aspect of the evaluation process is to get a handle on which areas will be most exposed to high winds, which areas will be flooded and how deeply, and which areas will be susceptible to waves as well as rising water. Coastal elevations are critical in this regard, and the initial and most important first approximation at hazard mapping can usually be done with the FEMA FIRMs, available at any town hall or nearby county courthouse or from FEMA (see appendix for address). These maps delineate V zones (100-year flood level plus waves of 3 feet or higher), A zones (100-year flood level), B zones (100- to 500-year flood level), and C zones (areas above the 500-year flood level). Probable extreme- and high-hazard zones are derived from the FIRMs (fig. 4.3). Very simply, areas within V zones will always be extreme hazard, and areas in A zones

will be high or moderate hazard depending on the amount of forest cover. Areas above these most floodable zones will be moderate to low risk, at this first level of rating, depending on the amount of forest cover. If hurricane inundation maps (SLOSH maps) are also available, potential overwash zones can be noted, adding detail for fine-tuning the initial risk assessment (fig. 4.4). The U.S. Geological Survey has produced posthurricane hydrologic survey maps for several hurricanes, and these maps are excellent sources for identifying future flood potential and expected storm-surge water elevations.

In general, U.S. Geological Survey topographic maps are of limited use in determining detailed land elevations on barrier islands because contour intervals are typically too large. The FIRMs must be field checked as there are errors on some of them, and the maps tend to be conservative in outlining the V and A zones. SLOSH maps of potential overwash can become quickly outdated if dunes have been removed in a recent storm or if new frontal dunes form either naturally or with the help of bulldozers and sand fencing. Likewise, removal of dunes for construction, whether legal or illegal, will increase risk from storm surge as well as wind.

The second step in the preliminary risk assessment is to evaluate vegetation type, distribution, and density. As mentioned previously, dense maritime forest can be invaluable in reducing exposure to wind and wind-borne

Table 4.7   Factors to Determine Potential for Property Damage in the Coastal Environment

---

Island Setting
  Morphology
    1. Lagoon width
    2. Lagoon depth
    3. Island width
    4. Island elevation
    5. Dunes
      a. Number of ridges and location
      b. Elevation and Width
      c. Volume
      d. Paths or road gaps found between or through sand dunes/vegetation
      e. Dune walkovers
    6. Inlets
      a. Present-day location
      b. Historical location
      c. Potential location
      d. Width
      e. Depth
      f. Behavior
    7. Shoreface slope (degrees)
    8. Erosion rate (m/yr)
    9. Shoreline orientation (degrees)
  Vegetation
    1. Types
    2. Native vs. nonnative
    3. Cut or cleared areas
  Offshore and Beach
    1. Sediment type and longshore current direction
    2. Rock type (if any)
    3. Slopes (degrees)
    4. Beach width
      a. Narrow
      b. Wide
    5. Presence or absence of sand dunes

Type, Location, and Density of Development
  Type of Development
    1. Commercial
      a. Type(s)
      b. Density
    2. Residential
      a. High- or low-rise
      b. Single-family home
      c. Density of development
    3. Utilities
      a. Type(s)
      b. Location and planned expansion
      c. Storm worthiness
  Location of Development
    1. Oceanfront
    2. Lagoon side
    3. Interior of island
    4. Evenly distributed
    5. Piers
      a. Built through dunes and vegetation
      b. Built on beach, extends into ocean
    6. Infrastructure
      a. Drainage for surface runoff and sewers
      b. Road's location relative to the shoreline
        -Perpendicular
        -Parallel
        -At an angle
    7. Other
  Density of Development
    1. Undeveloped
    2. Sparsely developed
    3. Partially developed
    4. Fully developed

Shoreline Engineering (Stabilization) Type and Location
    Seawalls
    Jetties
    Groins
    Breakwaters
    Others

Storm Response (Observations to be made after a storm in the context of developed vs. undeveloped areas)
  Beach/Dune Response
    1. Storm impacts on replenished beach
    2. Extent/amount of erosion
    3. Width of dry beach
    4. Effect on beach and offshore rocks if present
    5. Offshore or beach rock protection of development
  Seawall Response
    1. Performance of different types of seawalls
    2. Seawall effects on development behind wall
    3. Scour at base of seawalls
    4. Presence/absence of beach in front of wall
    5. Width of beach in front of wall
  Overwash Characteristics
    1. Overwash penetration and areal extent
    2. Overwash thickness
    3. Fate of overwashed sand (returned to beach?)
    4. Roads and passages that acted as overwash passes
  Inlet Changes
    1. Distance of study site from inlet
    2. Inlet changes
    3. Ebb and flood tidal delta effects
    4. Inlets that formed in areas narrowed by development

Table 4.7 continued

5. New inlets and effects on:
   a. Development
   b. Infrastructure (would burying utilities help?)
   c. Mainland

Storm-Surge Flood/Ebb Response
1. Storm-surge inundation category
2. Funneling of storm-surge return water
   a. Natural (e.g., low spots in dunes, low elevation areas)
   b. Funneled by development (e.g., roads, building constrictions, storm sewer outfalls)
3. Storm-surge ebb scour channels

Development Response
1. Any large buildings that provided "wave shadow"
2. Any small buildings that smashed into larger ones
3. Did different types of structures respond differently in terms of storm worthiness?

In a given situation, some of these factors may be of little importance, but this list can be a useful checklist to be certain a parameter has not been ignored.

4.3 Flood insurance rate maps (FIRMs) are available from the Federal Emergency Management Agency and from any city hall. These maps help in determining if your property is in a flood zone and, if so, how high your house must be elevated to comply with federal codes (and to avoid 100-year flood levels). This particular map shows a portion of the north coast of Puerto Rico, east of San Juan. See text for description of flood zones.

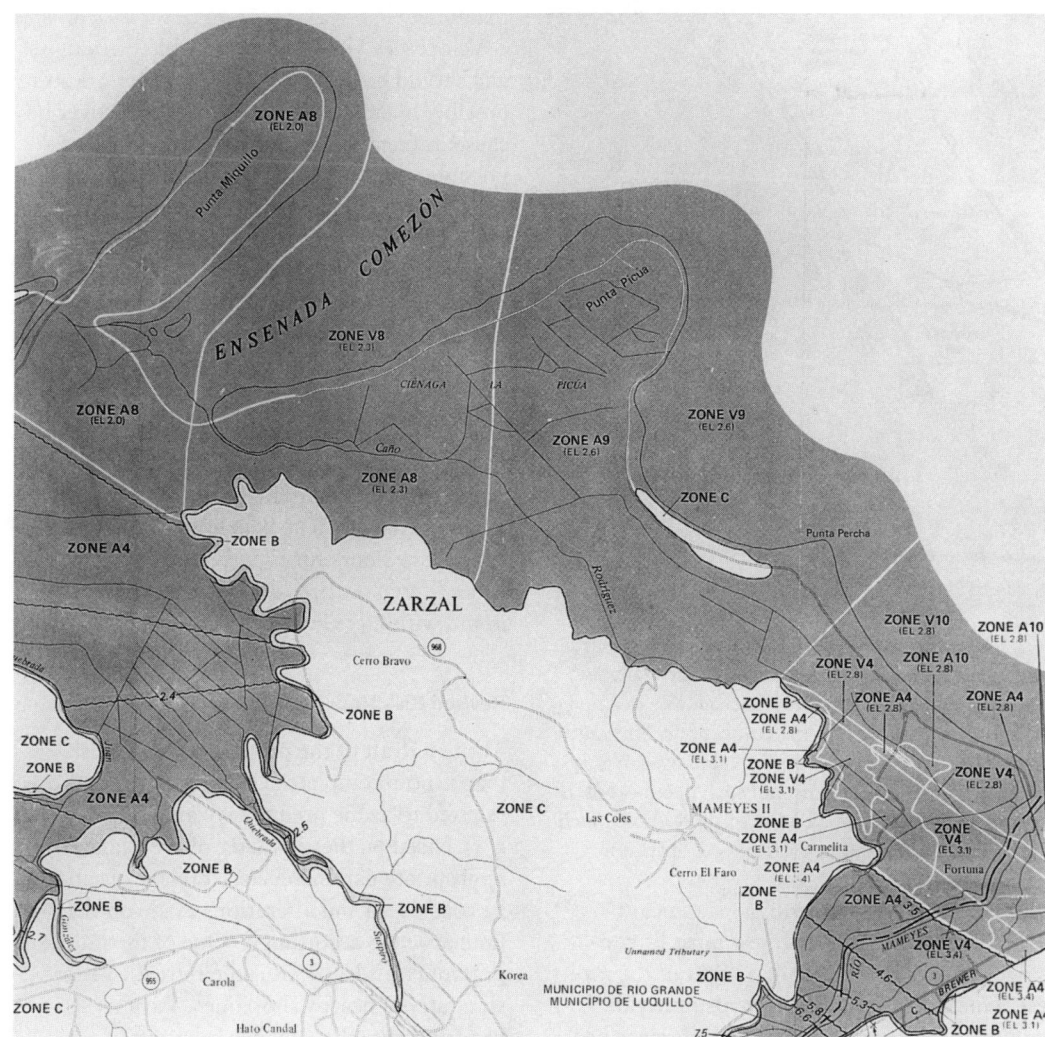

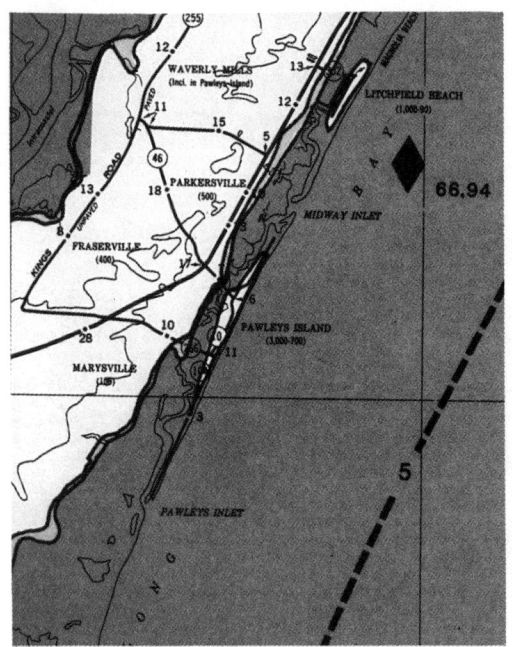

4.4 SLOSH map of Pawleys Island, SC. These maps are available from your state emergency management agency and are used to determine evacuation routes and for selecting shelter locations. For South Carolina, separate maps are available for category 1 through category 5 storms, but such detail is not available in every state. This map shows how much flooding would occur in a category 1 hurricane.

debris. Thus, any area with a "significant" cover of dense maritime forest must be considered lower risk than areas without the forest. Thick shrub thickets may also provide enough protection to make the difference be-

tween undamaged, damaged, and destroyed buildings.

Velocity (V zones) are in trouble, period, and should not be developed. A zones are still possibly high hazard even with forest cover because of their susceptibility to flooding and possible wave action. High-elevation areas with forest are lowest risk, but high-elevation areas without forest are considered moderate risk because of wind exposure. Low-elevation areas with forest are moderate risk, and low-elevation areas without forest (unfortunately often the most heavily developed areas) are at high risk.

Vegetation cover can be determined initially from aerial photographs or USGS orthoquads, but must be field checked for accuracy and ongoing changes. The field effort can require from a few hours for islands entirely in A or V zones to two or three days of mapping for an island with more complex topography.

Revised Risk Map

The first draft of the property damage risk map is now complete. The map so produced is referred to as the preliminary analysis map (fig. 4.5). Based on the character of the island, the appropriate parameters (see table 4.5) can now be considered and alterations made on the preliminary map accordingly. One of the most important considerations is the retreat rate (erosion rate) of nearby shorelines. Various historical shoreline-change maps often are

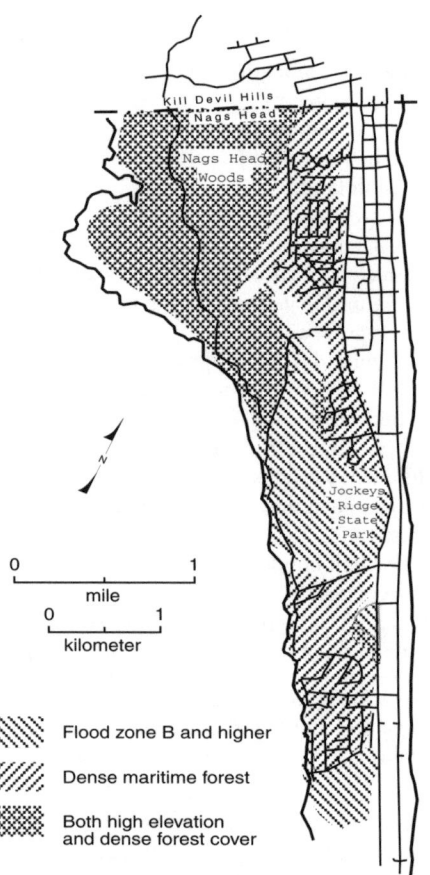

Flood zone B and higher

Dense maritime forest

Both high elevation and dense forest cover

4.5 Preliminary risk map of Nags Head, NC, based on the primary risk factors: elevation and forest. Areas with both high elevation and dense forest cover are at lower risk; areas within flood zones and without forest cover are at highest risk. Unshaded area equals V and A zones

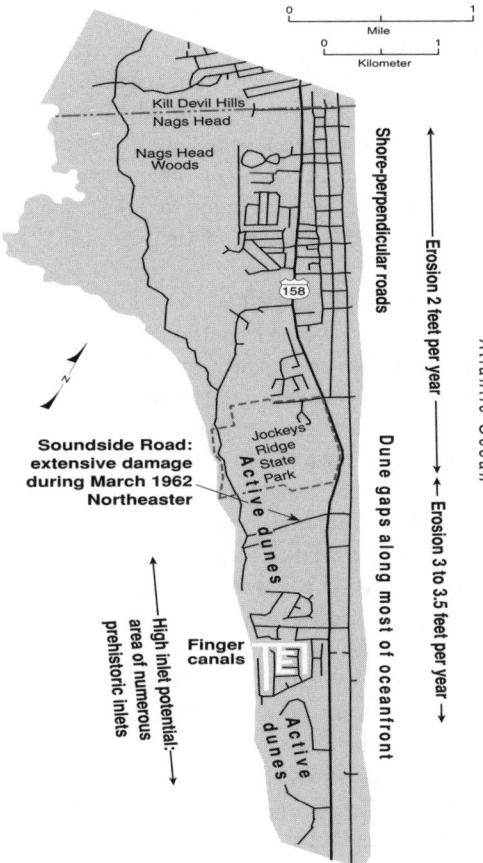

Map labels:
- Mile (scale 0–1)
- Kilometer (scale 0–1)
- Kill Devil Hills
- Nags Head
- Nags Head Woods
- 158
- Atlantic Ocean
- Shore-perpendicular roads
- Erosion 2 feet per year
- Erosion 3 to 3.5 feet per year
- Dune gaps along most of oceanfront
- Soundside Road: extensive damage during March 1962 Northeaster
- Jockeys Ridge State Park
- Active dunes
- High inlet potential: area of numerous prehistoric inlets
- Finger canals
- Active dunes

4.6 Secondary risk factors used to refine the preliminary coast risk map (fig. 4.5). These site-specific considerations include erosion rates, development patterns, historical storm response, and engineering projects. Each of these may have a positive *or* negative effect on the final risk ranking (see table 4.3).

available through state coastal management agencies or U.S. Army Corps of Engineers reports (see appendix for address). Clearly, if a moderate-risk zone is likely to be adjacent to the beach in a decade owing to high erosion rates, the hazard classification must be changed to high or extreme risk. Similarly, former inlet positions are mapped, past storm effects recorded, widths of dune fields measured, and patterns of good or poor development identified, allowing map revision. These secondary factors are summarized on the risk revision map (fig. 4.6), which is the basis for deriving the final ranking of the risk zones.

One of the more difficult parameters to evaluate is the quality of construction. A poorly constructed house or building truly is a "loose cannon" and will launch or float off its piers or foundation into adjacent houses during a storm. This battering-ram effect and missiling of debris from adjacent houses is a genuine hazard that increases the vulnerability of houses that may be well built. Strong building codes do not necessarily reduce the risk. Building codes are only as good as their enforcement. Newer isn't necessarily better, as newer designs are often less wind resistant. The hurry to build new houses can lead to shoddy building techniques. One of the most telling incidents comes from Hurricane Andrew (1992). Houses built by Habitat for Humanity stood nearly unscathed while neighboring houses were destroyed or heavily damaged: houses built with care, in part by amateurs,

stood better than houses built by professionals. The quality of adjacent construction is a factor in risk analysis. One loose cannon may sink your ship.

Another factor that's difficult to evaluate is the role of existing or planned engineering structures, either designated for shoreline stabilization or for navigation. In the case of existing structures, their performance in past storms can be evaluated. New structures are unknowns, but impacts may be predictable.

Other common questions that may need to be qualitatively addressed are:
—Is there a nearby jetty planned or in place that will increase the rate of erosion?
—Are the frontal dunes still present along the shoreline, or have gaps been cut by storms or by communities making vehicle beach access points?
—Are the frontal or interior dunes about to be removed (within a decade or so) by natural erosion or human activities (development)?
—Is there a likelihood of inlet migration away from or toward a particular site?
—Is new inlet formation likely or was there an inlet here in a previous storm that was filled (and is likely to form again)?
—Are there cross-island streets that may allow flooding or overwash into the interior?
—Do the map zones correspond well to known island response in previous large storms?
—Is there a wide beach storm buffer, natural

or replenished, that is likely to quickly disappear?
—Are there mobile homes nearby which can
be assumed to blow apart and furnish
wind-blown "missiles"?
—Are there nearby buildings, old and pre–
building code, that are also likely to be
destroyed and contribute to missiling or
ramrodding?

Using tables 4.2 to 4.6 as guides, each pro-
cess/feature that applies to the coastal region in
question is investigated and evaluated. Historical storm impacts, for instance, may be recorded in popular magazines, newspapers,
books, or garnered from "person on the
street" interviews. Talking with longtime residents can provide a wellspring of information;
however, wave heights and wind speeds often
seem to "grow" over the years with each retell-
ing of the story! The so-called gray literature
(e.g., agency reports, consultant reports, unpublished theses) and nonscience sources (e.g.,
tax maps, histories, newspaper accounts, community historical documents) can also provide
useful data and general insights.

All of this information is now considered for
incorporation in the final risk ranking (see fig.
4.7), occasionally resulting in reclassification
of zones to more hazardous categories than
those of the preliminary analysis that are based
only on elevation/vegetation considerations.
Although the initial draft of the community
risk map based on the FIRMS, SLOSH maps,
elevation, and vegetation is a semi-quantitative

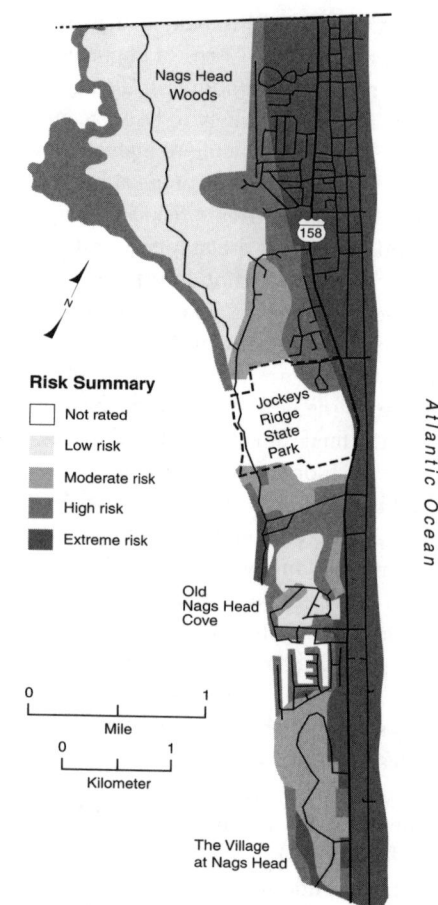

4.7 The final coastal risk map based on information
from figures 4.5 and 4.6. This map shows our judg-
ment of the relative risk to property from a head-on
hit by a moderate category 3 hurricane.

approach, incorporation of the numerous
other factors is largely subjective. The accuracy
of this input and the revision depends on the
quality of information (e.g., of inlet behavior,
response to previous storms) and the experience of the individuals doing the mapping. In
actual practice, we have found that the changes
from the preliminary analysis map are usually
minor.

The FIRM and SLOSH maps also have a certain qualitative aspect even though boundary
lines are drawn at specific elevations. Many
unknown factors control flood elevations and
overwash penetration on barrier islands.
Many, if not most, big storms have "unusual"
or "unexpected" aspects to them, resulting in
more intense flooding or wave penetration
overall or at any given location than predicted.
This imprecision is a result of both our lack
of complete understanding of all the physical
aspects of both barrier islands and storms, and
of random turbulence and the general chaotic
nature of storms. We can't predict the weather
with absolute precision, and we certainly are
unable to predict its effects any better.

On the other hand, our experience is that
the risk zones on barrier islands defined in
tables 4.1 and 4.4 are quite obvious and their
mapping fairly reproducible. Delineation of
risk zones, while fairly simple to those familiar
with island storm responses, may not be nearly
as obvious to individual homeowners or even
local elected officials and planners, yet with a
little effort the process of mapping can be understood.

Hazard mapping is ideally suited to the application of Geographic Information System (GIS) computer technology. Island physical and geomorphic (landform) descriptive criteria are entered into a computer data base, then, using GIS, any set of criteria can be combined to give the user the type of assessment desired. The preliminary analysis can be made by GIS summing of elevation and forest cover digitally entered as separate layers. The secondary factors for revising the preliminary map can be added, each as a separate digital layer, or summed separately depending on the user's needs.

GIS has four advantages over more traditional mapping systems:

(1) It is digital, making data handling and transfer simple.

(2) Maps can be produced depicting part or all of an island or community at any desired scale.

(3) New data can be entered into the database as it is developed.

(4) Updated maps can be easily produced.

For certain criteria (e.g., dune gaps, cement block houses, houses on grade versus on pilings), a house-by-house assessment of ranking criteria can be made while driving along a street and entering observations into a laptop computer.

Variable Scales

Based on the criteria just described, three hazard assessment scales are recommended:

—An *inter-island scale* compares relative overall risks between entire islands.

—An *intra-island scale* examines risks within portions of a single island.

—An *individual building scale* provides a site-specific ranking by building or small groups of buildings.

Inter-Island Scale

The inter-island scale can be thought of as a comparison of large-scale coastal segments, through which the overall risk factors for an entire island or shoreline reach are determined relative to other islands. The ranking is based on an overall geologic setting and coastal processes evaluation which considers property damage risk but ignores political boundaries and hazards to life. Many of the processes that affect barrier islands are large in reach or area (e.g., wave refraction patterns, longshore drift, and currents that may affect long stretches of shoreline on front, back, or inlet sides of an island) and are influenced by the overall island geometry (e.g., tidal deltas, beach/dune ridges, dune fields). This big-picture scale addresses island-length and cross-island processes relative to sediment supply, long-term changes in island morphology, and potential changes in plant distribution.

Figure 4.8 is an island-scale ranking of the North Carolina barrier islands based on natural characteristics. The ranking is on a 0–4 scale, 0 being poor (highest risk) and 4 good

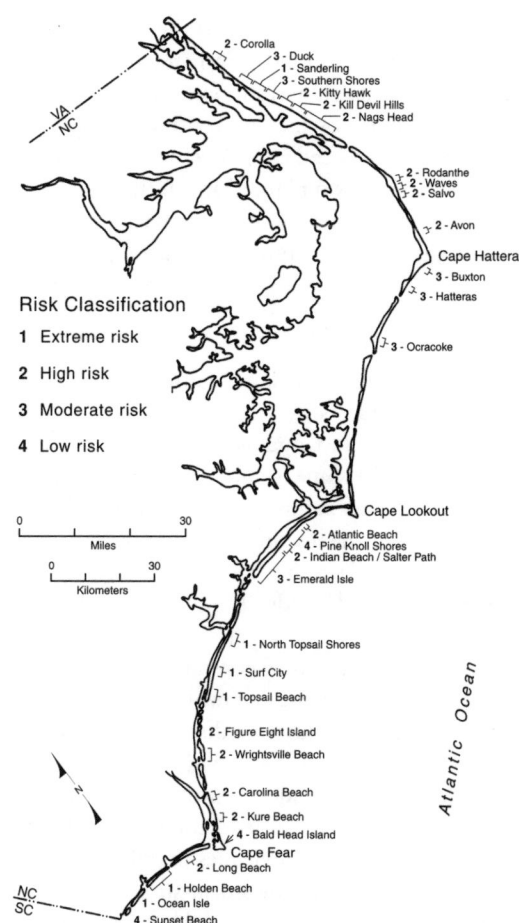

Risk Classification

1 Extreme risk

2 High risk

3 Moderate risk

4 Low risk

4.8 A generalized risk classification of all the islands in North Carolina. Clearly "islandwide" classification is painting with a broad brush since on most islands a range of risk category sites exist. For this classification, lower numbers mean higer risk.

(least risk). Table 4.8 expands the ranking, adding three additional parameters. This type of ranking is useful to pinpoint areas of concern and areas in which more study or resources may be needed to develop a property damage mitigation plan. Maps of this scale are also useful as a first level of evaluation for individuals seeking to purchase coastal property.

Intra-Island Scale

The second level of risk evaluation, the intra-island scale, ranks various hazard zones of one portion of an individual island relative to other portions of the same island. The risk map in figure 4.7 is an example of the intra-island scale map for the northern portion of Nags Head, North Carolina. The intra-island, or community, scale is ideally suited to application of GIS computer technology, and provides a basis for developing a community mitigation plan. At this scale one can evaluate features such as the continuity of a dune line or forest cover within a specific stand of forest. Service grids would be considered at this scale (e.g., Are streets running perpendicular to the shoreline? Are utility or water lines parallel to the shore likely to be cut by storm-surge processes?). Locating in the middle of an area designated extreme or high risk is not wise economically even if the specific site can be mitigated against hazards.

Table 4.8   Relative Island-Scale Risk Ranking of North Carolina Barrier Islands

	Natural Characteristics	Management Quality	People Danger	Crystal Ball	Total
Sunset Beach	4	3	2	3	12
Ocracoke	3	3	2	3	11
Nags Head	2	4	2	3	11
Figure Eight	2	3	2	3	10
Pine Knoll Shores	4	2	2	2	10
Emerald Isle	3	2	2	2	9
Buxton/Hatteras	3	2	2	2	9
Long Beach	2	2	2	2	8
Bald Head	4	1	1	2	8
Wrightsville	2	2	2	2	8
Indian Beach/Salter Path	2	2	2	2	8
Kitty Hawk	2	2	2	2	8
Duck/Southern Shores	3	2	2	1	8
Kill Devil Hills	2	1	2	2	7
Holden Beach	1	2	2	2	7
Kure/Carolina	2	1	2	1	6
Atlantic Beach	2	1	2	1	6
Avon/Salvo/Waves/Rodanthe	2	1	1	2	6
Topsail Beach	1	2	1	2	6
Ocean Isle	1	1	2	1	5
Surf City	1	2	1	1	5
Sanderling	1	2	1	1	5
Corolla	2	1	1	1	5
North Topsail Shores	1	0	0	1	2

This inter-island level ranking includes the evacuation hazard (People Danger) and also a subjective ranking of future development trends (Crystal Ball).
Scale= 0: highest risk to 4: lowest risk.
Risk Classification shown in figure 4.8 is based on island natural characteristics only.

4.9 The white house in the center of this photo is on a low-risk site. It is well elevated and is surrounded by dense maritime forest. The gently sloping roof is an aerodynamic design and should help deflect wind if the roof is properly secured to the walls. This house is in Salter Path, on Bogue Banks, NC. This and figures 4.10, 4.11, and 4.12 are examples of the *individual-building scale* of coastal hazard risk assessment.

4.10 This large, new house in Nags Head, NC, is at moderate risk to storm damage. The high elevation precludes storm-surge flooding and wave attack. Although wind exposure is high, a properly designed and constructed house may suffer little damage in a storm. Failings in the design of this particular house include its sharp, angular roof design, which will catch the wind in a storm and increase uplifting forces, and the lack of shutters, which will likely make it impossible to "protect the envelope" of the house.

Individual Building Scale

The third level of risk assessment, that of individual buildings or groups of buildings, involves more detailed description and assessment of sites and local alterations of the natural island environment. At this scale, building codes and compliance as well as architectural design and construction can be considered, if desired. Figures 4.9 through 4.12 are examples of this most detailed scale of assessment. This site-specific scale of assessment also lends itself to application of GIS technology, making on-site building-by-building assessments. Community planning maps (tax maps, zoning maps) can be used as possible base maps.

The Need for Updating

Barrier islands are dynamic features, and even in their native state, without human alteration, risk assessment of sites can change dramatically, either gradually over decades or rapidly in a single big storm. Shackleford Banks, North Carolina, on the Cape Lookout National Seashore, affords an example of such change. At the eastern end of this east-west–oriented island, large dune fields once existed. They were even used as lookout points to spot whales by the island's inhabitants in the 1800s. Then in the late nineteenth century, a series of hurricanes destroyed houses and discouraged agricultural pursuits. People disassembled the houses that remained and moved them to Harkers Island (a lower-risk island in Pamlico Sound) and the mainland. The settlements on

4.11 A lot for sale in Nags Head Pond subdivision of Nags Head, NC, is a high-risk location despite the presence of mainland forest. The low elevation (note the ponded rainwater) means the area will be susceptible to storm-surge flooding and direct wave attack during storms.

4.12 A multifamily unit in Nags Head, NC, on an extreme-hazard site; a narrow beach with no dunes and no forest offers no protection from storm waves.

the island have long since disappeared, the last inhabitants moving off after the 1899 hurricane. The dunes at the eastern end now are completely gone, having been lost to storms and rapid shoreline erosion, mostly within the last three decades. An area that would have been rated a moderate-risk zone for development has changed naturally to high or extreme risk as the island character evolved from a dune-dominated system to one impacted mainly by storm overwash. At about the same time, farther to the west, an active dune field developed and migrated over the maritime forest, significantly changing elevations and eliminating forest cover.

Equally significant changes can be made by humans. On Ocean Isle, North Carolina, an interior dune ridge, the only one on the island, was flattened to make way for development. The lowered elevation put the entire development in the A zone, leading to a change in risk classification from the preliminary to the final mapping from moderate risk to extreme risk. The change of the island topography didn't happen 50 years ago or even 25, but in 1994 (fig. 4.13)! That risk increase will be reflected immediately in property owners' insurance

4.13 On Ocean Isle, NC, the central dune ridge of the island was largely removed by bulldozing in 1994 for development. This ridge was the only portion of the island above the 100-year flood level, an important reason to prohibit the bulldozing of interior dunes.

costs because V zone properties pay the highest flood insurance rates. Remember: Most developers seek to maximize their profit, and this may not be in the purchaser's best interest with respect to potential property damage.

Another example of change, impacting on property damage risk, can be seen in Kitty Hawk, North Carolina. The dune along the shorefront once extended along the entire community. It was constructed in the 1930s by the Civilian Conservation Corps to halt the shoreline erosion and to provide a "protected"

4.14 Kitty Hawk, NC, was protected until recently by the dune constructed in the 1930s by the Park Service. Since the dune's erosion and complete breaching in the late 1980s, the first few blocks of the community have been flooded several times, as in this example in the 1991 Halloween storm. It is not possible to replace the frontal dune because the beach has retreated past the former location of the dune. A new dune could be constructed on top of the frontal road, which could be a very effective storm buffer, but blocking the road would not be popular among dwellers and businesses. Photo by Rob Young.

4.15 The jetties shown here at the south end of Ocean City, MD, were constructed after a new inlet opened during a hurricane in 1933. Sand trapped by the jetties caused sand starvation and migration of Assateague Island (at the bottom of the photo). Now Assateague's open-ocean shoreline lies landward of its 1933 lagoon shoreline, a spectacular example of island migration.

4.16 The Morris Island Lighthouse now stands hundreds of yards at sea, a victim of shoreline retreat that was much hastened as the Charleston jetties halted the sand flow. The shoreline migrated past the lighthouse in the mid-1940s.

area along which to build a road. The artificial dune failed to change erosion rates, but it did afford some protection for buildings. After 50 years of dancing, however, the piper now demands payment. During the 1980s, the dune began to deteriorate due to storm penetration, and the 1991 Halloween northeaster finished the job by creating large gaps in the dune, resulting in flooding of portions of the community. The dune cannot be rebuilt in place because the old dune location is now occupied by the beach, backed up against the frontal road. Between the time of dune construction and 1991, the community had only experienced major flooding once, in the great 1962 Ash Wednesday storm. Since the 1991 storm, the community has been flooded four times (fig. 4.14)!

The effect of shoreline engineering on a whole-island system is starkly portrayed in figure 4.15. It has taken several decades to be fully realized, but the impact of the Ocean City, Maryland, jetties on the next island to the south, Assateague Island, Virginia, can now be seen, over 50 years after the jetties were constructed. Similar stories abound along the coast. The Morris Island lighthouse, once on the back side of Morris Island, now stands some 2,000 feet at sea, a sentinel that watched Morris Island rapidly migrate away after the Charleston jetties were built in 1898 (fig. 4.16).

Risk mapping must be viewed as a dynamic process. Risk maps should be reexamined at least on a decadal time frame and after each major storm. Even without storms, dune systems can change, shorelines can retreat, forests can be killed or removed. Major storms are bound to make changes on the island, and of course they are the ultimate test of the validity of the risk maps. Meaningful mitigation must be ongoing, adaptable to natural and human-induced changes, and applied islandwide, not just along the island front. Figures 4.17 and 4.18 are concept diagrams, detailed flow-charts illustrating the several reasoning steps, sources of information, and relationship of parameters that go into coastal risk mapping. Cultural and historical responses to natural processes change, requiring periodic revision of the risk maps.

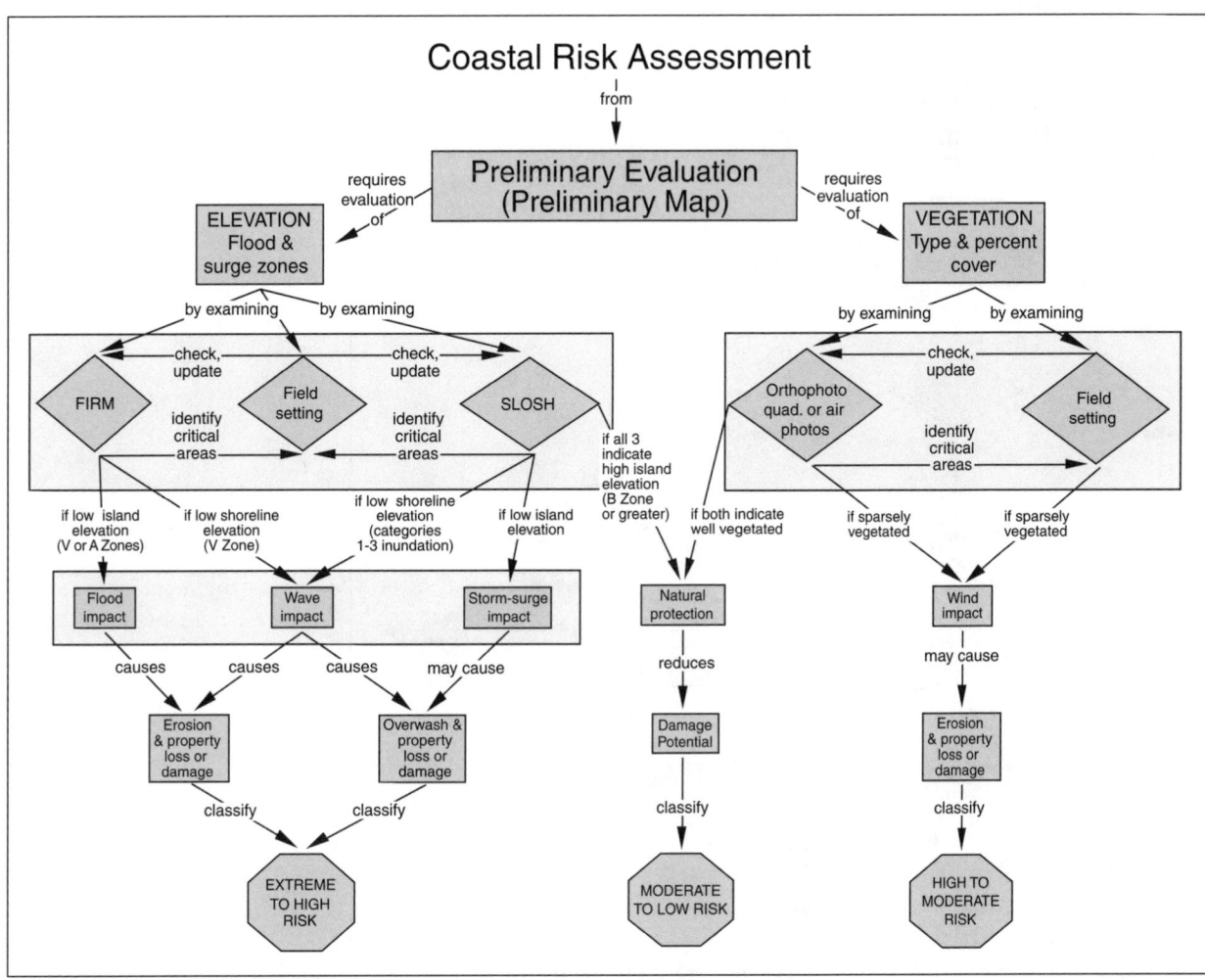

Coastal Risk Assessment

from

Preliminary Evaluation (Preliminary Map)

requires evaluation of

ELEVATION Flood & surge zones

requires evaluation of

VEGETATION Type & percent cover

by examining — by examining

check, update

FIRM

Field setting

SLOSH

identify critical areas — identify critical areas

by examining — by examining

check, update

Orthophoto quad. or air photos

Field setting

identify critical areas

if low island elevation (V or A Zones)

if low shoreline elevation (V Zone)

if low shoreline elevation (categories 1-3 inundation)

if low island elevation

if all 3 indicate high island elevation (B Zone or greater)

if both indicate well vegetated

if sparsely vegetated

if sparsely vegetated

Flood impact

Wave impact

Storm-surge impact

Natural protection

Wind impact

causes — causes — causes — may cause

reduces

may cause

Erosion & property loss or damage

Overwash & property loss or damage

Damage Potential

Erosion & property loss or damage

classify — classify

classify

classify

EXTREME TO HIGH RISK

MODERATE TO LOW RISK

HIGH TO MODERATE RISK

4.17 Concept diagram of the preliminary phase of the coastal risk assessment mapping procedure and steps leading to risk classification. Boxed information includes parameters and processes evaluated, and diamonds enclose specific databases examined and field checks. Preliminary evaluation leads to initial risk assessment, which is modified according to the steps in Figure 4.18

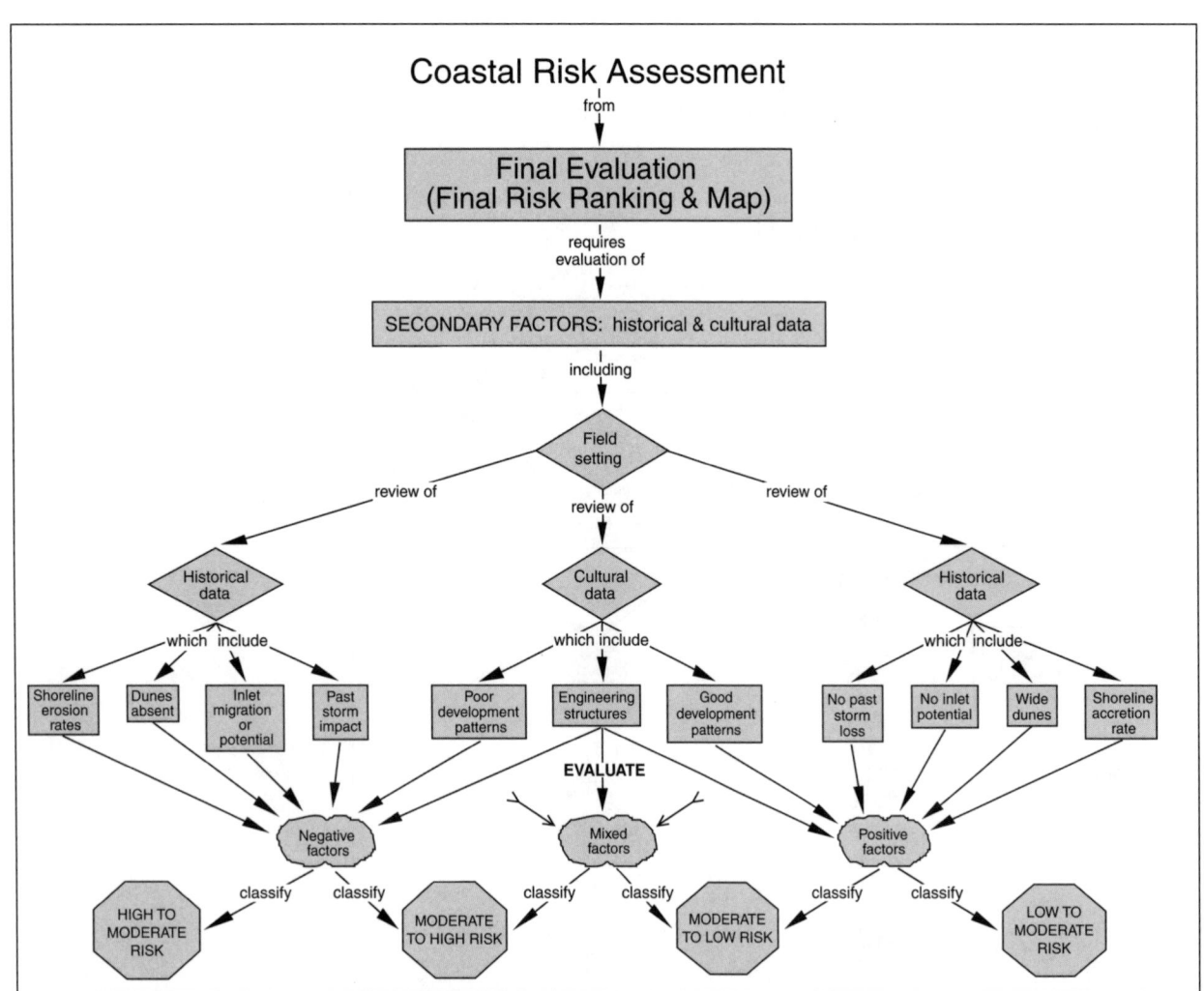

4.18 Concept diagram of the final evaluation phase of the coastal risk assessment mapping procedure and steps leading to risk classification. Boxed information includes parameters and processes evaluated, and diamonds enclose specific databases examined and field checks. Final evaluation is based on the preliminary evaluation as presented in Figure 4.17 and includes evaluation of secondary factors leading to the final risk map.

5 Evolving Approaches to Property Damage Mitigation: Focusing on the Island Front through Engineering and Regulation

The human perspective of coastal hazards tends to be biased, focused on the sea and the shoreline. The erosive waves roll into the front of the island. The furious storm approaches from offshore with winds and storm surge, as does the tidal wave. Shoreline erosion is marked by scarped dunes, toppled trees, decks and houses collapsed on the back of the beach, and streets that suddenly end in the space above the beach. To prevent or reduce such damage, the various mitigation schemes and techniques that evolved tended to concentrate on the shoreline (fig. 5.1). This shoreline focus was also driven by the early development patterns that concentrated buildings on the ocean front and the fact that coastal protection was the charge of a specific governmental agency, the U.S. Army Corps of Engineers (see USACOE, 1984). If the history of property damage mitigation were an adventure film, the story line would go something like the following:

—Shoreville is built on the shoreline.
—Shoreville gets in trouble (storms/erosion).
—Rescue: shoreline engineering (hold that line).
—Shoreville's beaches disappear: more trouble.
—Rescue: setbacks and zoning (keep 'em off the line).
—Time passes and the shoreline catches up to the setback.
—Rescue: the new (old) engineering: beach nourishment.
—More beach erosion, more property damage, more financial loss.

—Shoreville continues searching for a solution.

The question is: Can there be a happy ending for Shoreville? The answer depends on how the town continues to respond to the natural hazards. We will evaluate and discuss several "Shorevilles" and their various approaches to coastal zone management in this chapter and in chapters 6, 7, and 9. Good, bad, right, or wrong, each example gives insight to living by the rules of the sea!

Our society has learned much in the past several decades about coastal processes and about the natural changes that occur on beaches and coasts. Our attention was drawn first to the beach, where we observed that retreating shorelines may undermine houses, destroy dunes, and bring the surf zone closer to the buildings, increasing the storm damage potential. Traditional coastal property damage mitigation techniques have basically fallen into two broad categories: engineering and land-use planning (Godschalk, Brower, and Beatley, 1989). Engineering is used to alter the landscape and/or strengthen buildings. Such mitigation efforts have concentrated on the beach, either armoring the shoreline, replenishing the beach itself, or moving buildings back from the beach, largely neglecting the near-beach interior areas. Land-use planning is used to control the locations of buildings and thereby reduce (or limit) the number of buildings and people at risk.

Victims of History?

When America's first barrier island resort communities developed, no grand plan was followed. The focus was on the recreational beach, sea bathing, and taking in the air. Hotels were constructed at the back of the beach and summer boardwalks laid down over the dunes to be taken up at the end of the season. As the crowds increased, so did the number of structures, including dance and amusement pavilions and cottage communities. The boardwalks became fixed, and so did the communities. Elsewhere along the U.S. shore, barrier island towns grew from a variety of beginnings, including old plantation summer retreats, fishing camps, church camps, and hunting reserves, always driven by one common element: access. The way over the water from mainland to island may have been a nineteenth-century causeway, an early railroad or carriage bridge, but the access provided the initiative for growth. All of these communities shared one thing: None of them knew the nature of the islands on which they were located.

In a matter of decades, most of these towns and local developments began to experience island migration, what they perceived as beach erosion, thinking that it was just a small stretch of beach that was moving just a little bit, not realizing or understanding that the islands were actually responding to a rising sea level. The mindset of "defending" the shoreline or "reclaiming" the land prevailed. What started with a few logs or rocks to protect a

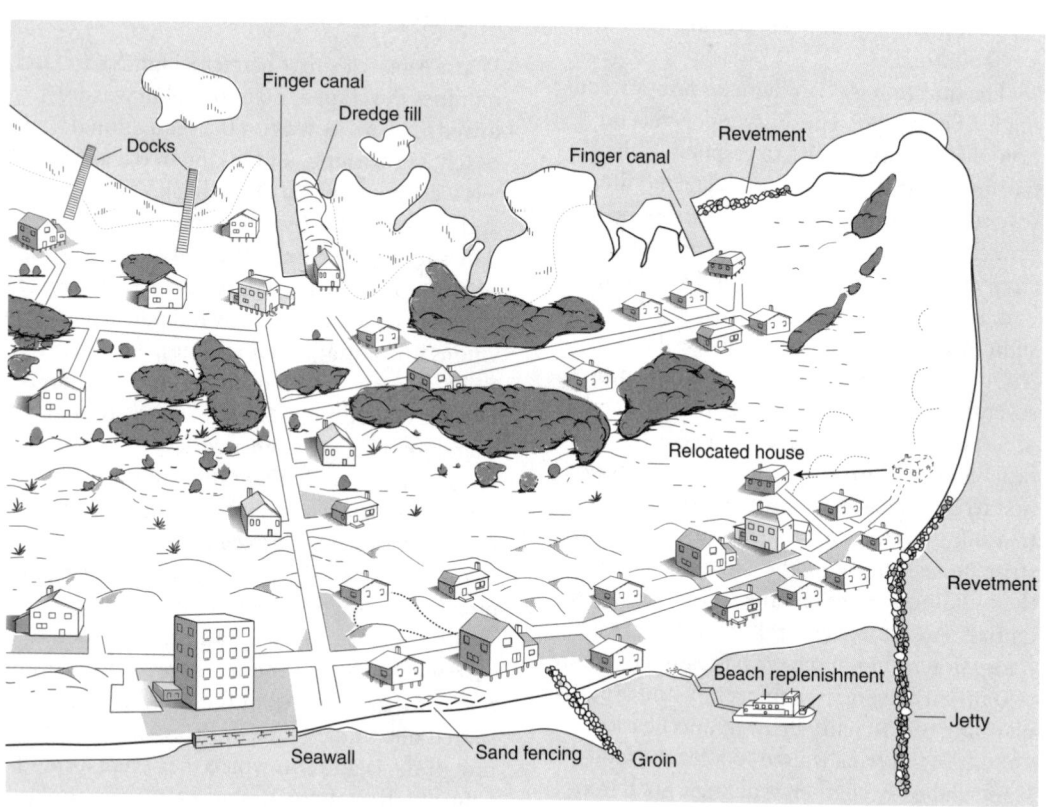

Finger canal
Dredge fill
Docks
Finger canal
Revetment
Relocated house
Revetment
Beach replenishment
Jetty
Seawall
Sand fencing
Groin

5.1 Pandora's Island with various typical frontal property damage mitigation structures and procedures. The pluses and minuses of these approaches are apparent. Groins and jetties trap sand (moving from left to right in this example) and widen the beach, but downdrift beaches erode. Seawalls cause the loss of the beach in front of them but protect buildings. Beach replenishment is costly and temporary.

single house grew to small walls, then larger walls, revetments, and massive engineering works and beach replenishment projects. The experience also was perceived as a local problem, not of regional or global extent. As a result, the patterns of access and new development continued into the post–World War II era of affluence, mobility, and the continued

fashionability of the seashore—each new community in turn impacted by storms or the ultimate end of the day-to-day dynamics of the migrating barrier island.

The Great Age of Engineering

Although settlements like Diamond City, North Carolina, and Edingsville Beach, South Carolina, threw in the towel and moved after devastating hurricanes in the 1890s, few communities even considered relocating as an option. The technology to move out of harm's way existed. In 1888 the large, wooden Brighton Beach Hotel on Coney Island, New York, was moved back from the shoreline using steam locomotives (fig. 5.2). But relocation was not given its due in a time devoted to great engineering marvels. This was the era of the Brooklyn Bridge, the Eiffel Tower, the beginning of plans for the Panama Canal. In a time when no river was too wide to bridge, dam, or jetty, no isthmus too great to breach with a canal, the natural response to a disaster like the Galveston Hurricane of 1900, which left over 6,000 dead, was to build a mighty protective wall against such future events (fig. 5.3). It was humans against the elements, and no one doubted that humans could out-engineer the forces of nature.

The Rivers and Harbors Act of 1889 charged the U.S. Army Corps of Engineers with maintaining navigable waterways. Maintaining inlet and river mouth entrances to har-

5.2 The Brighton Beach Hotel on Coney Island, NY, was one of the first buildings to be moved back from an eroding shoreline. The hotel was moved back 2,000 feet (600 meters) in 1888. Photo from *Scientific American*, April 14, 1988.

bors was achieved through the construction of great jetty systems (e.g., Nantucket Harbor, Massachusetts; Charleston Harbor, South Carolina; Savannah Harbor, Georgia; Fernandina and Jacksonville Harbors, Florida). Developed ocean shorelines were also the shores of navigable waters, so it followed that the Corps would combat erosion on the fronts of developed islands using the arsenal of engineering (solid structures) to defend against nature. Coastal engineering, the building of seawalls and groins, became the pattern for property damage mitigation. These structures shared one thing: directly or indirectly they contributed to the loss of the beach that fronted the structure (the raison d'être for most coastal communities).

The coming of the automobile and post–World War II affluence spurred on barrier island development, but our understanding of island dynamics had not yet come of age. Hurricanes and erosive storms likewise continued, as did the archaic engineering "solutions." Seawalls and groin fields remained the common options available to communities in trouble due to shoreline retreat, storm-surge flooding, and frequent overwash. The seawalls that were growing larger with time may have given a false sense of security with respect to the protection (or lack thereof) afforded against hurricanes and northeasters. In many communities, cottages were replaced by larger housing units (i.e., duplexes, "sea cabins," condominiums, high-rise complexes).

5.3 In 1900, 6,000 people died when a hurricane crossed Galveston Island, TX. One response of the community was to construct a massive seawall, shown here in 1961 (a) before and (b) after Hurricane Carla passed by. Photos supplied by Marcus Milling.

The Sea Bright, New Jersey, Example

Sea Bright, New Jersey, has been suggested as an example of the end result of following the philosophy of hard stabilization. Pilkey and Wright (1988) analyzed historic photographs and sketches to reconstruct the sequence of shoreline engineering and changes that took place in the Sea Bright area. Their work indicates that the area was an undeveloped spit (barrier sandbar attached to the mainland) into the nineteenth century. By 1868 the settlement of Nauvoo, a cluster of fishermen's shacks and sheds, was spread over the low ground behind the natural, unstabilized beach, which was steep and narrow.

The first "permanent" house was built in 1869 at the north end of the development, and by 1877 the shoreline was lined with a row of houses at the back of the beach. These buildings were positioned near the high-tide line with little or no dune protection. Almost immediately they were in trouble, and by 1886 protective walls had been built in front of some of the houses (fig. 5.4). A postcard dated 1903 shows a large rubble wall lining the beach at the north end. This may be the rubble-mound wall installed in 1898 (Kraus and Pilkey, 1988). In 1931 the rubble-mound wall was similar to the present wall in terms of dimensions, but a recreational beach remained in front of the wall. The remains of this beach are held in place today by the long groins in front of the three- to four-block stretch of "downtown" Sea Bright. But in either direc-

5.4 This is the oldest known photo of Seabright, NJ. It was taken from the deck of the French ocean liner, *L'Amerique*, which ran aground in 1877 in fog. In those days, Seabright was a lightly settled, low-lying barrier spit with no form of shoreline stabilization. Photo kindly furnished by George Moss.
5.5 Seabright, NJ, today is a walled city. This community, which is the closest open-ocean New Jersey beach community to the New York metropolis, is considered to be the endpoint of "Newjersey-ization." This is a term referring to a long walled and beachless shoreline.

tion of the groins virtually no beach remains in front of the seawall (fig. 5.5).

The original first row of houses also is gone and today the high seawall (17 foot or 5.2 meter crest elevation) blocks the view of the sea. The massive wall "protects" an area of less than one square mile and a population of fewer than 2,000 year-round residents. In 1984 a northeaster, possibly a 30-year storm, caused $82 million in damages, primarily to the seawall and beaches. This amount was essentially equal to the value of all of the buildings in town. What would the cost be in a really big storm?

Sea Bright truly is at an end point. Maintaining the wall is more costly than the value of the property it is supposed to protect (the wall potentially can trap flood waters and contribute to property damage). The beach, robbed of its local sand supply by the seawall and probably impacted by redistributed wave energy off the wall, has all but disappeared. The community does not have the financial resources to maintain the wall and must go hat-in-hand to the taxpayers at the federal and state level for coastal "welfare." A more prudent management alternative would be the gradual removal and/or relocation of the buildings. In the short term the community must look for ways to mitigate damages other than reliance on the seawall.

The Shift from Engineering to Regulation

Initially, armoring the shoreline to hold it in place and protect beachfront buildings and property through engineering was the method of choice. Beach replenishment is a modern equivalent of the old engineering "fix" to hold the line. More recent methods have been through regulation, often loosely put into the category of land-use planning. Land-use planning and zoning efforts have taken a broader approach and attempted to consider the entire island or coastal zone. Looking at the high density of development along much of our shores, we can easily see that land-use planning either hasn't worked or is a Johnny-come-lately to the arena of coastal zone management. Table 5.1 is a list of options that have been utilized by communities and individuals to reduce the potential for coastal property damage (also see fig. 5.1). Examples include stricter building requirements for buildings in flood zones, especially velocity zones (i.e., the island front) where the flood level is topped by waves, and setback requirements for houses built on the eroding fronts of islands. Hazard-sensitive zoning requirements also often focus on the front sides of islands.

By the 1950s and 1960s the realization that coastal buildings were subjected to higher winds and flooding (even those behind seawalls) led many states and communities to adopt more stringent building codes to strengthen buildings in the coastal zone. By the 1970s the national experience dictated that

Table 5.1 Beachfront Property Damage Mitigation Options

Abandonment

Relocation
  Active (relocate before damaged)
  Passive (rebuild destroyed structures elsewhere)
  Long-term relocation plans for communities

Soft Stabilization
  Adding sand to beach
    Beach replenishment
    Beach bulldozing/scraping
  Increasing sand dune volume
    Sand fencing
    Raise frontal dune elevation
    Plug dune gaps
  Vegetation
    Stabilize dunes (oceanside)
    Plant marsh (soundside)

Hard Stabilization
  Shore parallel
    Seawalls
    Bulkheads
    Revetments
    Offshore breakwaters
  Shore perpendicular
    Groins
    Jetties

Modification of Development and Infrastructure
  Retrofit houses
  Elevate houses
  Curve and elevate roads
  Block roads terminating in dune gaps
  Move utility and service lines into interior or bury below erosion level

Zoning, Land-Use Planning
  Recognize hazard areas and avoid:
    Tidal inlets (past, present, and future)
    Swashes
    Permanent overwash passes
  Setbacks
  Choose elevated building sites
  Lower density development

Things to Keep in Mind:
  Each island or coastal community is different.
  Consider the entire coastal zone, not just oceanfront.
  Rising sea level must be considered.

---

something be done to control the losses incurred from hurricanes and great storms like the Ash Wednesday storm of 1962. The tremendous loss of habitat (salt marshes, shell fisheries) also was being recognized as fisheries were forever lost or closed down due to pollution. The response was twofold: the National Flood Insurance Act of 1968 (also the result of the tremendous loss of property on riverine floodplains) and the Coastal Zone Management Act of 1972. Building requirements were upgraded, and many coastal states began to define critical environments and control development through permit processes. Communities and states adopted approaches such as zoning and setback requirements.

Land-use planning control is, in truth, a before-the-fact approach. In other words, it provides guidelines that are easy to incorporate into newly developing areas but difficult in already developed (that is, most) coastal areas.

An example is the requirement that all new structures be built a minimum distance back from the shore. Such *setback regulations,* however, are only interim or temporary "solutions" to the erosion problem. Setback laws require buildings to be positioned a defined "safe" distance from the shoreline, this distance usually being determined by the natural rate of erosion. In North Carolina, for example, oceanfront single-family houses must be set back 30 times the average annual erosion rate, while larger structures such as hotels and condominiums must be set back 60 times the average annual erosion rate. The average annual erosion rate is determined from aerial photograph analysis of historical shoreline changes by a consultant to the North Carolina Division of Coastal Management. Choice of the magic number 30 has much to do with the life of a mortgage and little relationship to long-term survival of property. Thus, setbacks help protect the banker, but not the second generation of owners. Setback laws vary by state and community and may change over time. Check with your state or local coastal management office or town hall to find out your situation.

Zoning and land-use planning in the coastal zone and related issues are covered well in various studies. See, for example, Godschalk, Brower, and Beatley (1989) and Beatly, Brower, and Schwab (1994), as well as additional sources of information listed in the appendix. Although moving in the right direction, the regulatory approach has not stopped the rising cost of storm damage or the concentration of larger populations and their developments in these hazardous zones. Given the likelihood that few people living on barrier islands are going to get up and move to higher ground, society must look for ways to defuse the bomb.

Next Step in Property Damage Mitigation: PAR for the Course

In order to reduce the potential impact of coastal processes a more holistic approach to mitigation is needed. Society must move from engineering against nature to working with nature; from a focus on site-specific and linear (island front) regulation to a whole-island perspective; and from shore-hardening/hold-the-line programs to approaches that concentrate on preservation, augmentation, and repair (PAR) of the natural systems that we occupy with our development.

Coastal processes involve the broader "coastal zone," not just the beach and shoreline. Mitigation plans must take a whole-island point of view, including the nature of the coastal zone of the adjacent mainland. This view implies considering "nontraditional" mitigation approaches that pay attention to island interiors and lagoonal shorelines as well as the ocean beach and inlet shores. Coastal and island dynamics must be more completely incorporated in order to delineate existing and potential inlet hazard zones (see chapter 7) and to identify problem areas that may be evolving over the coming decades (e.g., loss of forest cover and destabilization or loss of dune fields). And just as these inner island areas must be given attention, the offshore, submerged island platform must be managed as part of the total system (e.g., tidal deltas and shoals may provide important protection from waves and a natural sand supply, so that their loss will have serious repercussions for the island system). As communities become more desperate for sand supplies, all eyes are turning seaward. The removal of offshore sand bodies will create a new set of problems for island property.

Nevertheless, the front side of barrier islands is a good place to begin a consideration of property damage mitigation.

Hard Stabilization: Hold That Line

Hard shoreline stabilization generally involves structures that either block and dissipate wave energy or that trap sand to widen a beach. A wide variety of types and designs of hard stabilization exists, but there are basically three major categories: (1) shore-parallel structures on land, such as seawalls; (2) shore-parallel structures offshore, such as breakwaters; and (3) shore-perpendicular structures, such as groins and jetties. Hard stabilization has been a very common shoreline management tool through the centuries, even before Roman times. In the

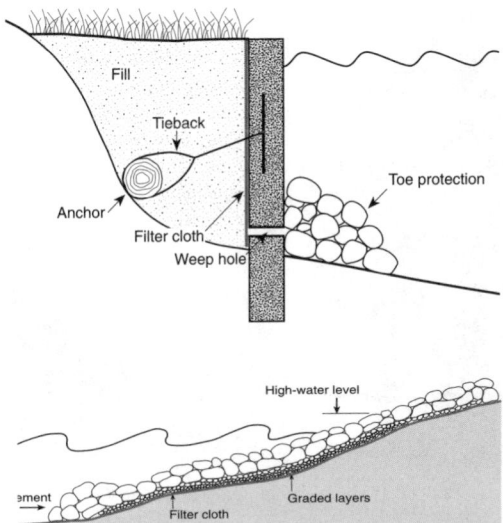

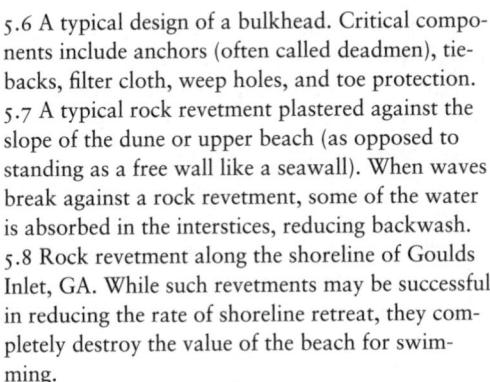

5.6 A typical design of a bulkhead. Critical components include anchors (often called deadmen), tiebacks, filter cloth, weep holes, and toe protection.
5.7 A typical rock revetment plastered against the slope of the dune or upper beach (as opposed to standing as a free wall like a seawall). When waves break against a rock revetment, some of the water is absorbed in the interstices, reducing backwash.
5.8 Rock revetment along the shoreline of Goulds Inlet, GA. While such revetments may be successful in reducing the rate of shoreline retreat, they completely destroy the value of the beach for swimming.

United States, many of the disadvantages associated with hard stabilization are only now understood after more than a century of use. See USACOE (1984) for a complete treatment of shoreline stabilization.

Shore-Parallel Structures on Land

Seawalls and their cousins are shoreline engineering structures that are built parallel to the beach on the subaerial beach. These structures are the most common type of hard stabilization.

Seawalls are wood, steel, rock, or concrete structures designed to protect the upland from the impact of waves (see fig. 5.5). Commonly, especially along the southern and mid-Atlantic

U.S. shoreline, seawalls are built at low elevation and are not intended to block storm waves, especially those accompanied by a storm surge. Such walls instead function mainly to halt the retreat of the shoreline into the line of buildings. Overtopping of seawalls is common in hurricanes, and walls fail for a variety of reasons, including build up of water pressure (from storm surge) on the landward side, storm-surge ebb flow tearing the wall apart, scouring and undermining by waves and storm-surge ebb, and direct wave attack. Seawalls should extend vertically well down into the beach to prevent undermining.

Bulkheads are generally indistinguishable from seawalls to the general public. In theory,

5.9 Trash revetment on the beach along the shore-line of Teller, AK, an Inuit village near Nome. Backing the World War II surplus debris is a line of 50-gallon drums.

5.10 A sandbag seawall in South Nags Head, NC. Such walls are legal in North Carolina, but the intent of the law is to provide temporary protection while a building is being moved. Sandbag seawalls do not have foundations and are susceptible to undermining and collapse in storms.

the primary purpose of bulkheads is to hold back the land from slumping or eroding into the sea and not to absorb wave energy (fig. 5.6). In reality, bulkheads serve both purposes. Usually the term refers to small, low seawalls.

Revetments can be a relatively inexpensive type of shoreline engineering. These structures consist of an armor of rock facing on a dune or beach slope (figs. 5.7 and 5.8). Their role is to act as a buffer to the waves, just like a seawall. As a wave breaks on a revetment, much of the water contained in the wave is absorbed in the interstices between the rocks, reducing erosion-causing backwash. In most storms, however, the difference between a revetment and a seawall is negligible. Often revetments are among the ugliest structures, especially those made of construction debris. Some are constructed of automobile transmissions, kitchen sinks, and old washing machines (fig. 5.9). Proper design requires carefully placed, heavy, wave-resistant material for the structure, properly angled and backed by filter cloth. Filter cloth is a decay-resistant mesh fabric that allows water to escape but prevents soil loss.

Sandbags are often used as a temporary erosion measure, but such walls rarely have a foundation, so they are swept away or moved about by the first storm (fig. 5.10). Bags, whether filled with sand or concrete, are best viewed as temporary protection, buying time before moving a building back.

Gabions also are used to construct seawalls or revetments. This type of structure consists of rock-filled rectangular steel wire mesh cages piled one on another to form a wall (fig. 5.11). Although the wire is plastic coated, the mesh inevitably corrodes, the gabion ruptures, the rocks spill out, and the wall disappears! Jagged wire and the rock leakage are strewn over the beach to form a hazard to beach strollers.

Table 5.2   Advantages and Disadvantages of Hard Shoreline Stabilization

Advantages
  Temporarily protects property
  May temporarily prevent shoreline retreat
  Low maintenance cost if properly constructed

Disadvantages
  Eventually causes loss of the recreational beach
    (erosion due to wave reflection/refraction;
    cuts off sediment supply)
  Increases erosion at ends of wall and/or
    downdrift
  Limits access to beach
  Often ugly (loss of aesthetics)
  Can be costly (up to thousands of dollars per
    foot)
  Requires regular maintenance (additional cost)
  Debris from walls become hazardous objects on
    beach or to structures in back of wall during
    storms

Properly designed, hard shoreline structures are dependable methods of halting shoreline retreat and protecting coastal property. Their disadvantages, however, are many, and are not solely aesthetic. Hard shoreline stabilization leads to degradation of the recreational beach, is costly both in the short term and long term, destroys beach aesthetics, makes beach access difficult, and is dangerous (table 5.2).

Seawalls lead to narrowing of recreational beaches by means referred to as active, passive, and placement beach loss. *Passive beach loss*

occurs in a landward-retreating shoreline situation, where a seawall (or highway or building) acts as a fixed reference point. As the shoreline continues to erode or retreat landward, the beach must narrow and eventually disappear. This process often takes several decades to complete. *Active beach loss* occurs when the seawall reflects wave energy or in other ways intensifies the surf-zone processes leading to beach sand loss. Little is understood about this mechanism as yet because of the difficulty in measuring and observing wave/current processes when most of the "action" takes place during storms. *Placement beach loss* occurs when a seawall is built seaward of the dunes and part of the beach is claimed by the seawall (figs. 5.12 and 5.13). That is, what used to be part of the recreational beach is now behind the seawall. Such placement of seawalls has led to beach loss in many locations, perhaps the most famous of which was Miami Beach prior to its 1981 replenishment.

All types of seawalls are subject to numerous hydraulic forces that must be accounted for in wall design and construction materials. Walls intended to prevent wave attack on buildings must be constructed high enough to prevent storm wave overtopping, but they also must be implanted deep enough in the beach to prevent undercutting. In general, the maximum depth of expected scour is roughly equal to the highest breaking wave at the site. Thus, maximum storm waves of 1 meter require a footing depth of at least 1 meter. Some scour will still occur, and placing large rocks at the foot of the struc-

5.11 A gabion seawall along the Aguada shoreline, west coast of Puerto Rico. Gabions are commonly used to construct seawalls along low to moderate wave energy shorelines. Gabions are wire baskets filled with cobbles. Their problem is that the wire often corrodes in seawater. Gabions probably last about five years on average before succumbing to corrosion or breakage.

ture will help reduce this effect, but it will also further reduce the recreational value of the beach.

A seawall or bulkhead protects only the land and buildings behind it. The ends should be joined to neighboring structures if possible. Where none exist, wing walls or tie-ins to the adjacent land must be built to prevent wave flanking by wave erosion at the ends of the wall. Wing walls are only a temporary measure because erosion will continue and extend beyond each successive flank built to "solve" the problem.

5.12 The Sandbridge, VA, seawall, where drainage of storm wave and rainwater has removed some of the fill in spite of the filter cloth. The tie-back cables leading to the anchors (see fig. 5.6) are exposed here. Construction on this wall was begun in 1989. Probably no other state would have allowed construction of this wall!

5.13 A group of students observe the damaged Sandbridge, VA, seawall in March 1992. The wall may have been lost because of weak tie-backs, poor foundation (the base was in mud), and lack of filter cloth. This wall caused placement loss of the beach when constructed because it was built seaward of the high-tide line. On the day the wall was completed, in 1989, the recreational beach was significantly narrowed.

Additional strength can be gained through the use of tie-backs (see figs. 5.6 and 5.12). Tie-backs anchor the upper part of the wall with steel cables or rods to logs or other anchors (called deadmen) embedded deep into the beach or bluff. Accumulating water and soil pressure will build behind bulkheads and seawalls. Drainage must be provided to allow water to escape from the landward side of the wall without carrying the backfill material with it. Backfilling with gravel and having frequent openings (weep holes) along the lower part of the wall allow water to escape. Walls should be backed by filter cloth to prevent the backfill from washing out.

This brief discussion of seawalls is not intended to give design advice, but rather to illustrate some of the considerations of construction and indicate why these walls fail. A

professional engineer should be consulted prior to any construction. Seawalls and other types of shoreline engineering structures require maintenance. Remember to ask the engineer how much maintenance will cost, for how long the design is "guaranteed," how the wall will perform under typical storm conditions (to be expected during the wall's design life), and what the impact of the wall will be on the adjacent property and beaches.

Environmental Impact of Seawalls

A seawall doesn't absorb all of a wave's energy—some of that energy moves off the wall by reflection, scouring, and eroding sediment in front of the wall. Some of the energy is also deflected along the wall to the adjacent unprotected property where the energy is spent eroding the shore. In addition, the narrowed beaches in front of seawalls lead to a reduced sand supply to adjacent beaches, further exacerbating erosion of neighboring beaches. Where wall-protected areas form miniheadlands, the waves may be refracted by the wall into the adjacent unwalled shore, again concentrating erosive wave energy. Stroll along the shore of a hotel or condominium row almost anywhere and observe where the beach is narrow or missing—the offending walls, bulkheads, or revetments will be obvious. The name of the community is of no consequence, the effects of seawalls are the same.

Unwise Building Location: Major Structure

America's shoreline is dotted with examples of buildings *initially* located too close to the shore. Folly Beach, South Carolina, provides a convenient example. Folly Island is developed along its entire length and width, with the exception of its extreme ends. The northern end is a U.S. Coast Guard facility, the southern end is a county park. Wood-frame, single-family beach cottages and several small commercial buildings compose the majority of the island's development. The dominant structure on the island is the Holiday Inn, located at the end of S.C. Route 171. The Inn was built in 1985 on the site of the old Folly Beach Pavilion, a popular gathering place since before World War II. The Folly Beach Pavilion/pier was opened in the 1920s and burned in 1957. It was rebuilt in

1960 but burned again in 1977 and was rebuilt in 1995. When it was first opened, the pavilion had a wide, healthy beach in front of it. By the time it burned the second time, the high-tide recreational beach was gone.

As mentioned, the Holiday Inn was built in a very dangerous position, essentially out on the beach (although elevated)! The recreational beach would not exist, not even at very low tide, if the beachfront were not nourished. This island is in such an erosive state that even though a replenished beach was emplaced in the spring of 1993, there was no high-tide beach by spring of 1994, even without a major winter storm! Locating this building on the shoreline was a mistake that should not have been allowed to happen and would have been easy to avoid.

Seawalls, Sediment Loss, and Narrowing Beaches

The immovable object has another significant effect. Beaches often get part of their sediment from erosion of the land at the back of the beach. Walls cover up that source, starving the beach of its sediment supply, either in the area of the wall or downdrift of the wall. Such sediment starvation is a particular problem in places along the New England and Long Island, New York, shoreline where erosion of old glacial sediments is a major source of beach sands and gravels.

If beaches are the attraction, and we build walls to protect property at the expense of the

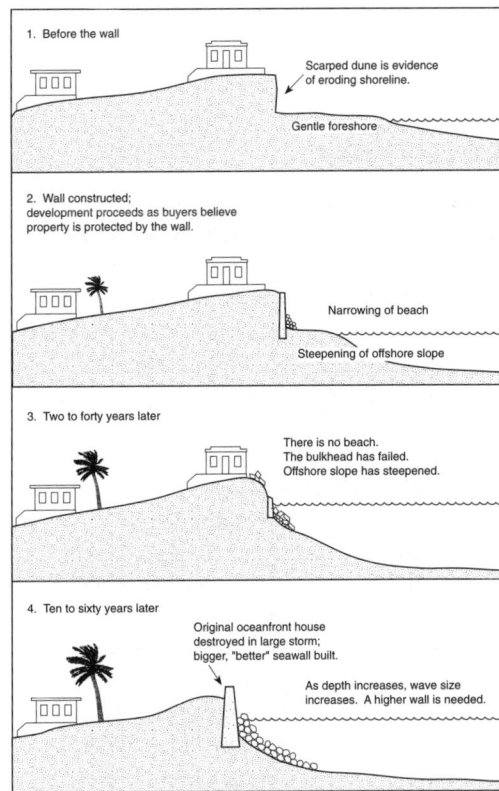

5.14 The seawall saga. Seawalls, built to halt a retreating shoreline, provide a static object against which the beach narrows and may disappear altogether. The process may take several decades; by the time the Seabright, NJ, stage is reached (fig. 5.5), fragments of broken walls litter the zone in front of the wall.

beaches' sand supply, then we kill the goose that laid the golden egg (fig. 5.14).

Lengthening Walls and Narrowing Beaches

Regional comparisons of dry beach (recreational beach) widths in front of seawalls to dry beach widths on beaches not altered by engineering structures indicate that beaches are consistently narrower in front of seawalls. In New Jersey the narrowest average dry beach widths are found in front of seawalls; the widest beaches were unstructured (Hall and Pilkey, 1991). This relationship is not unique to New Jersey, however. In a study comparing developed stabilized and unstabilized beaches along the East Coast (Pilkey and Wright, 1988), dry beach width is consistently and significantly narrower in front of seawalls. That same study showed that stabilized beaches in the San Juan, Puerto Rico, metropolitan area averaged only about 3 to 7 feet (1 to 2 meters) in width, while unstabilized beaches along the same stretch averaged 60 feet (over 18 meters) wide.

Offshore Breakwaters

Breakwaters are offshore structures, usually shore parallel, specifically designed to reduce wave energy and shelter a portion of the shoreline. The effect of breaking wave energy is to interrupt or lessen longshore transport of

sand, causing accumulation of sand behind the structure (figs. 5.15 and 5.16). As the local beach widens, this starves downdrift beaches of sand, similar to the effects of groins and jetties discussed below. In the 1980s and 1990s breakwaters again became "fashionable" with proponents of shore hardening, and new breakwater systems consisting of portable concrete segments were installed in Florida and New Jersey. Although it is necessary to observe shoreline structures on a decadal time frame to

5.15 Offshore breakwaters off Colonial Beach, VA, in Chesapeake Bay. These breakwaters have created a wave shadow, causing sand to be trapped behind them, and have successfully built a swimming beach. As is apparent here, the breakwater can also create a significant swimming hazard.

5.16 The Reddington Shores, FL (west coast), breakwater was quite successful in trapping sand, but in a little more than one year, five swimmers were killed on the rocks. The trapped sand may be leading to the narrowing downdrift beach (visible at the top of the photo).

understand their impact, early reports indicate these "portable" breakwaters are not succeeding as claimed by their "inventors." At the same time warnings were being raised about the ill effects of these new structures, Palm Beach, Florida, was reversing the trend. The Palm Beach breakwater, built in the early 1990s, was scheduled to be removed by summer 1995. The device created more erosion problems than it solved, including the area in back of the breakwater.

Shore-Perpendicular Structures

Whereas shore-parallel structures are built basically to block wave energy, shore-perpendicular structures are designed to block the alongshore flow of sand. The trapped sand, in theory, is held as a beach deposit.

5.17 Diagram showing the impact of groins. Groins are walls that are perpendicular to the shoreline; they can be constructed out of almost anything. They trap sand on the updrift side, causing erosion on the downdrift side. In general, groins are a net loss for beach quality and erosion rates on most shorelines.

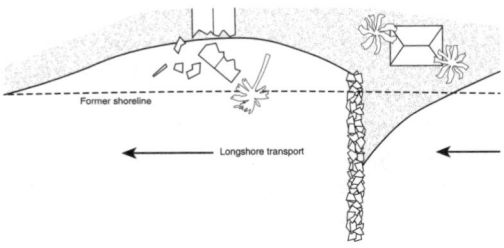

Groins are walls, built perpendicular to the shoreline, designed to trap and hold sediment flowing in the longshore current, thus rebuilding an eroding beach (fig. 5.17). Made of rock, wood, concrete, or steel, groins are also useful in retention of sediment already on the beach or retaining new sand from a nourishment project. Groins are often, but not always, used in conjunction with seawalls and sometimes with beach replenishment. Groins can be low or high, long or short, depending on the desired extent of sand trapping. Spacing of groins in a "groin field" depends on local sand supply and wave characteristics and is highly variable from one beach to another.

The problem with groins is that they trap sediment on one side and intensify erosion on

the other, depending on the net littoral drift direction. Updrift beaches are widened, downdrift beaches are "starved" of sediment. Many lawsuits have been initiated by property owners claiming increased erosion damage as a result of nearby groin construction.

Groins should not be used where there is low or no supply of sand flowing laterally along the beach. When no sand builds up, the groins are not doing their job. After a long period of time, or very quickly during a storm, the shoreline can retreat past the landward end of groins, causing them to become detached from the shoreline. Once detached, water and sediment pass between the groins and the beach, rendering the groins useless (fig. 5.18).

5.18 On Dauphin Island, AL, shoreline retreat has stranded out at sea the groins that were once "protecting" the beach. There are many other examples of this phenomenon in the United States, including the shoreline at the Cape May, NJ, Coast Guard station.

The bottom line on a worldwide basis, according to some engineers and geologists, is that groins are a losing proposition for beaches. Sand buildup temporarily gained in one place simply transfers the erosion problem to another place and worsens the overall erosion problem.

Jetties are analogous to groins, but constructed specifically to "stabilize" navigational inlets and entrances. The structures are intended to make navigation safer and channel maintenance cheaper. Jetties cause more extreme downdrift erosion than groins because they more completely interrupt the longshore sediment transport system. The great number of jetties along America's shores, particularly the Atlantic and Gulf Coasts, suggests a widespread and significant frequency of associated erosion problems. Along the east coast of Florida, jetties may be the principal cause of shoreline retreat. Jetties that extend far out to sea may also channel sand offshore during storms, causing a permanent loss of the sand supply in the longshore drift. Newer jetty designs call for sand bypassing systems to the sand in the longshore drift. Bypassing systems, however, cannot operate during storms, the time when much of nature's transport work is being accomplished. Jetties are the antithesis of sand conservation in an island system.

Jetties are necessary to provide safe harbor entrances, but they protrude so far out into the nearshore sediment transport system that they can essentially block all the sediment from be-

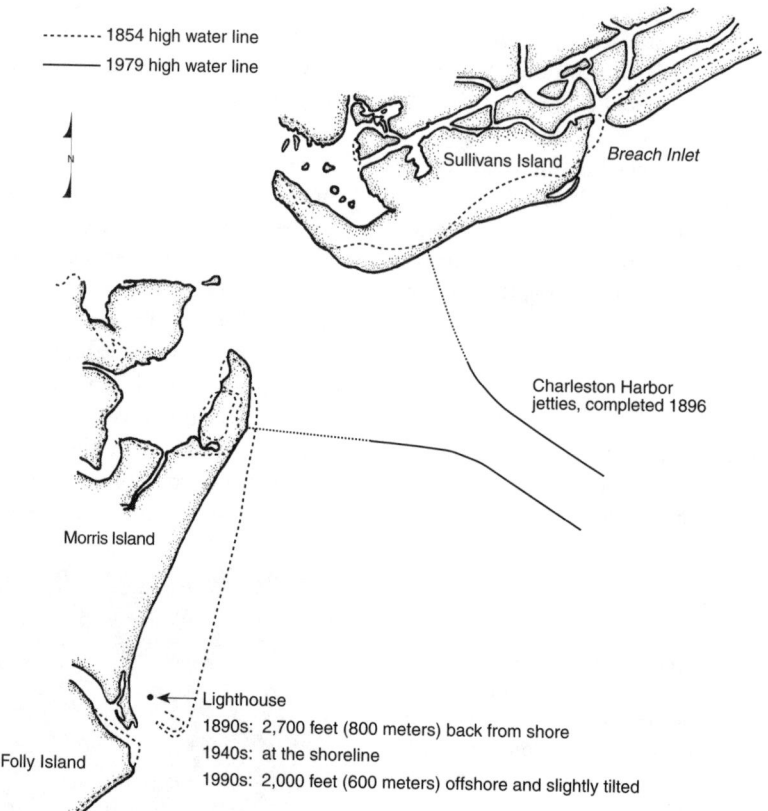

ing transported along the coast, causing "sand starvation" on the downdrift side. The resulting erosion can be spectacular. Such a starved situation exists on Morris Island and Folly Island, South Carolina (fig. 5.19). The two islands sit in the sand transport "shadow" of the Charleston Harbor jetties. Sand that

5.19 A map of the entrance to Charleston Harbor showing the jetties that were completed 100 years ago. Largely as a result of the jetties, Morris Island, the real-world site of the events chronicled in the movie *Glory,* has migrated a considerable distance. The lighthouse, once on the back side of the island, is now at sea. (Adapted from Zarillo, Ward, and Hayes, 1985.)

would normally travel southward along the coast is trapped by the jetties, resulting in severe erosion of both Morris Island and Folly Island. The jetties were built in the 1890s. A 1935 Army Corps of Engineers report to Congress (House Document 156, 74th Congress) indicates that erosion rates on Folly Island were 7 feet (2.1 meters) per year for the island, and as high as 51 feet (15.5 meters) per year at Stono Inlet. The present-day erosion rates vary widely from about 1.5 to 6 feet (0.5 to 1.8 meters) per year according to studies done by the South Carolina Coastal Council. The Corps recommended then, as now, hard stabilization to combat erosion.

The long-term effects of the Charleston jetties are illustrated in figure 5.19. When the jetties were completed in 1896, the Charleston Lighthouse on Morris Island (also referred to as the Morris Island Lighthouse) stood some 2,700 feet (more than 800 meters) inland. With the jetties cutting off the natural sand transport, starving Morris Island of sand, the shoreline began eroding rapidly and by 1940 had eroded back to the lighthouse. The lighthouse today stands, slightly tilted, some 2,000 feet (600 meters) offshore (see fig. 4.16). Total shoreline retreat: 4,700 feet (1,430 meters) in 90 years—over 50 feet (almost 15 meters) per year! The story is just the opposite on Sullivans Island and Isle of Palms, South Carolina, located north (updrift) of the jetties. There the beach has widened over the years.

A similar situation has occurred with the jetties at Ocean City, Maryland (see fig. 4.15). The jetties blocked the north-south sand transport, causing updrift accretion, but the entire island of Assateague Island, Virginia, has migrated landward by a distance equal to its width in less than 50 years.

The time frame of these occurrences is the point that needs emphasis here. After the Charleston jetties were built, it took over 40 years for the shoreline to erode back to the lighthouse and another 50 years to reach the present situation. In less than 50 years Assateague Island migrated back one full island width. The full impact of massive shoreline engineering projects cannot be ascertained within just a few months or even a few years of limited data and observations. Several decades are needed in order to fully understand the system. Unfortunately, most people cannot think (or plan) in a decades-long time frame, leaving our children and grandchildren to be the ultimate witnesses and victims of the impact of projects being designed today.

One example is the proposed jetty system at Oregon Inlet, North Carolina. The proposed jetties of over one mile in length will block sand transport to Pea Island, the Cape Hatteras National Seashore, and all the villages therein. A National Park Service panel of distinguished scientists concluded that the proposed sand-bypassing system would not prevent such sand loss, but the jetty proposal marched forward. The reason for the jetties is to protect navigation into the inlet and reduce maintenance costs—all in support of a diminishing fishery of overfished waters. In a real sense, the current generation is stealing America's future beaches. In this case the theft is subsidized by taxpayers so that a few can make a final profit on the last of another resource. The rule (after the biologist Hardin) is: You can't do just one thing. Every action in nature has multiple effects.

Hard Stabilization: A Final Word

Once thought to be the ultimate solution to shoreline erosion problems, hard stabilization has increasingly gone out of favor. Seawalls have been the "solution" of choice for many years. The story of Cape May, New Jersey, has been retold often throughout the Living with the Shore series. In 1801 Cape May was the first full-fledged popular beach resort in America and sparked the beginning of the American rush to the shore. During the next 75 years, six presidents of the United States vacationed at Cape May (seven at Long Branch, New Jersey). At the time of the Civil War it was certainly the country's most prestigious beach resort. The resort's prestige continued into the twentieth century. In 1908 Henry Ford raced his newest-model cars on Cape May beaches.

5.20 (a) Miami Beach (circa 1972) has no beach along most of its length. The beach had not been lost due to beach erosion: too-far-seaward placement of seawalls had taken it away. (b) By 1981, the largest ($5 million per mile) replenished beach in the United States had been emplaced. The beach has lasted for almost 14 years, a record, but its hardness (which may be the reason for its long life span) prevents turtle nesting.

Today, Cape May is no longer found on anyone's list of great beach resorts. The problem is not that the resort is too old-fashioned but that little beach remains. Seawalls and groin fields were built to try to "save" the beach, but they did not slow the relentless encroachment of the sea. The long-term results are plain to see (see fig. 5.5). No beach remained until a recent replenishment project brought it back—temporarily.

Few question the thinking that if you want to protect buildings, massive seawalls will do the job. But do we want to protect only the buildings and not the beach? It is obvious that there are advantages as well as disadvantages to hard shoreline stabilization (see table 5.2). The question now is: Which are more important, beaches or buildings?

Soft Stabilization

Where "hard" stabilization is the construction of "permanent" structures, such as seawalls, to hold the shore in place and protect upland property, "soft" stabilization implies shoreline maintenance through the addition of new sand to replace the eroding beach, or planting vegetation to hold sediment in place. The goal is to protect property and maintain the economic and environmental value of the beach. The

approach is to work with natural processes rather than confront nature.

Beach Replenishment

Artificial nourishment is the modern method of maintaining a healthy beach to help protect buildings as well as provide a recreational resource. Beach replenishment involves placing new sand, from some outside source, on the beach. Reconstructing the beach is usually carried out by dredging, but sometimes dump trucks are used to bring in sand (e.g., in Virginia Beach, Virginia, and after Hurricane Hugo in North Myrtle Beach, South Carolina). Sources of sand are the continental shelf, inlets and associated tidal deltas, lagoons (not commonly used anymore because of improper sand quality and environmental problems), and inland sand pits. Beach replenishment is the most important, though not the only, form of soft stabilization. Replenishment appears to be the "wave of the future" in all states with heavy shoreline development. On the East Coast of the United States more than 100 beaches have been replenished since 1965, about 40 beaches have been replenished along the Gulf of Mexico shoreline, and 30 along the California shoreline (table 5.3; Leonard, Clayton, and Pilkey, 1990; Pilkey and Clayton, 1989; Dixon and Pilkey, 1991; Clayton, 1991; Pilkey, 1988). Miami Beach, the largest and most successful of all, cost about $5 million per mile for 10.5 miles (17 kilometers) of beach (fig. 5.20).

There are strong regional differences in the durability (lifespan) of replenished beaches, but there are some figures that communities can use to guess how long their new beach will last: on the U.S. East Coast barrier island shoreline (where the most information is available), replenished beaches, from Cape Canaveral, Florida, to the south, last on the order of nine years; between Cape Canaveral and the Florida state line, five-year lifespans are typical; between Florida and New Jersey, two to four years of replenished beach life can be expected; and along southern New Jersey, a two-year lifespan is more or less typical.

In most cases, replenished beaches erode much faster than the natural beaches that preceded them because the "new" beach is just piled on top of the upper beach, resulting in a beach that is steeper and out of equilibrium with the wave climate. Many of the standard

design parameters used to predict the durability (lifespan) of artificial beaches (such as grain size and beach length) seem to have little to do with the length of life of replenished East Coast beaches. Instead, storms and storm frequency are the most important factors. It is fair to say that current design methods for replenished beaches have failed to predict lifespans and nourishment intervals (the time span between nourishment projects). Experience on the East Coast has shown that the best way to predict how well a beach will last is to pump sand up and watch it! The first try will provide a gauge for the second and the third tries. Beach nourishment is not a permanent fix. In other words, despite all the mathematical and physical models, high-technology designs, and predictions, each individual beach replenishment project is really just an experiment—a very high-priced experiment.

The cost of replenishing open beaches by pumping sand from offshore can usually be assumed to be on the order of a bare minimum of $1 million per mile of shoreline per nourishment project. One million cubic yards per mile is a moderate-sized replenishment project, and dredged sand costs anywhere from $2 to $12 per cubic yard, depending on many factors. The estimate of the 50-year cost of replenishing a 36-mile (58-kilometer) reach of New Jersey beaches is $3 billion! Replenishing about 5 miles (8 kilometers) of Folly Beach, South Carolina, in 1993 cost around $12 million. The nourishment interval for Folly Beach was

predicted to be eight years, but based on previous experience in similar areas it probably should have been only one or two years. And, in fact, most of the dry replenished beach disappeared from Folly Beach by 1995—and without the passage of a significant storm! The 1990 projected cost for all South Carolina beach nourishment requirements was $65 million, which was a highly optimistic estimate based on a ten-year nourishment interval. Clearly, beach replenishment is a costly procedure.

Unfortunately, many of the costs of a beach replenishment project are hidden or not stated clearly at the beginning of a project, when the community is deciding if beach nourishment is the best alternative. In addition to the per-cubic-yard of sand cost, there are project design costs (which, depending on project size and design-phase model testing, can be over 20 percent of the total) and legal costs incurred (e.g., property condemnations). Also, the true cost of a replenishment project can be partially disguised as part of a cost/benefit ratio. For example, you might be told that replenishment will cost $5 million but the benefit (to tourism and storm damage reduction) will be $10 million for a cost/benefit ratio of 1/2 or 0.5, an almost impossibly good ratio. The method for arriving at the worth of the "benefits" is often questionable, however, and since the same people who are promoting the project are usually the ones calculating the cost/benefit ratio, they may make it appear much more favorable

Table 5.3 Partial Listing of Replenished Beaches on U.S. East Coast and Gulf of Mexico Barrier Islands

U.S. EAST COAST	Seaside Heights	*Delaware*	Hunting Island	Ft. Lauderdale By-	Longboat Key
	Seaside Park	Ft. Miles-Indian River	Folly Beach	the-Sea	Anna Marice
New York	Berkeley Township	Inlet	Seabrook Island	Hillsboro Beach	Mullet Key
Saganopack Pond	South Seaside Park	Indian River Beach	Hilton Head Island	Hillsboro Inlet	St. Petersburg Beach
Macox Bay	Barnegat Light	Beach Cove		Port Everglades	Upham Beach
Southampton Beach	Long Beach Island	Bethany Beach	*Georgia*	John U. Lloyd Park	Treasure Island
Great South Beach	Harvey Cedars		Tybee Beach	Hallandale Beach	Madeira Beach
Westhampton Beach	Surf City	*Maryland*	Sea Island	Hollywood-Hallandale	Redington Beach
Westhampton Dunes	Ship Bottom	Ocean City		Hanover Park	Indian Rocks Beach
Brookhaven and Islip	Brant Beach		*East Florida*	Bar Harbor	Clearwater Beach
Jones Beach	Union Township	*Virginia*	Mayport Naval Station	Miami Beach	Mexico Beach
Oak Beach	Island Heights	Virginia Beach	Jacksonville Beach	Key Biscayne	Okaloosa County
Gilgo-Cedar Beach	Long Beach	Sandbridge	St. Augustine Beach	Virginia Key	St. Joseph Spit
Lido Beach	Beach Haven		Brevard County		Panama City Beach
Rockaway Beach	Brigantine	*North Carolina*	Cape Canaveral	**GULF OF MEXICO**	Santa Rosa Island
	Atlantic City	Cape Hatteras	Indiatlantic Melbourne		Perdido Key
New Jersey	Ocean City	Fort Macon	Sebastian Inlet	*West Florida*	
Sandy Hook	Ludlum Beach Island	Atlantic Beach	Indian River County	Marco Island	*Mississippi*
Seabright	Upper Township	New River Inlet	Vero Beach	Keewaydin Island	Harrison County
Monmouth	Strathmere	Topsail Beach	Ft. Pierce	Naples	Bay St. Louis
Longbranch	Sea Isle City	Figure Eight Island	Lions Club Beach Park	Vanderbilt Beach	
Deal	Avalon	Wrightsville Beach	Jupiter Island	Bonita Beach	*Louisiana*
Shark River Inlet	Stone Harbor	Carolina Beach	Palm Beach	Ft. Meyers Beach	Isles Dernieres
Avon	North Wildwood	Bald Head Island	South Lake Worth	Captiva Island	Grand Isle
Belmar	Lower Township	Long Beach	Inlet	South Seas Plantation	
Spring Lake	Cape May		Delray Beach	Gasparilla Island	*Texas*
Sea Girt	Cape May Point	*South Carolina*	Boca Raton	Port Charlotte Beach	Galveston
Bay Head		Myrtle Beach	Pompano Beach	Venice Beach	Corpus Christi
Lavallette		Edisto Beach		Lido Key	South Padre Island

New England and Great Lakes beaches are not included.
Source: Pilkey and Clayton, 1989; Dixon and Pilkey, 1991.

than it actually will be. Be wary of cost/benefit ratios less than 1!

Beach replenishment can also be costly in an environmental sense. Dredging or pumping sand from offshore seems like a quick and simple solution to replace lost beach sand; however, such operations must be considered with great care. The offshore dredge hole may allow larger waves to attack the adjacent beach. Offshore sand may be finer in grain size or may be of a carbonate component that breaks up under wave abrasion. In all of these cases, the new beach will erode faster than the original. Dredging may also create turbidity that will kill bottom organisms. We believe the dredging of sand off Boca Raton for their new beach released mud that was responsible for killing coral heads and serpulid worm reefs more than 20 miles (32 kilometers) to the north. Streams of turbid waters from the surf zone of Miami Beach are still responsible for killing nearby coral heads 14 years after the beach was emplaced. Offshore, protective reefs may also be damaged by increased turbidity. Loss of reefs means faster beach erosion, as well as the obvious loss of fishery habitat and recreational diving locales.

The impact of beach replenishment on sea turtle nesting remains unclear. Because replenished beaches erode rapidly, they frequently exhibit an erosion scarp (small bluff on the upper beach of 1 to 4 feet—0.3 to 1.2 meters—or more), which prevents turtles from moving ashore. Such scarps are an obstacle for older or handicapped human beach users as well: easy to get down, difficult to get up. Miami Beach, probably because it is made up of irregularly shaped grains of seashell and coral fragments, is too hard packed for a turtle to excavate for a nest.

There is another kind of environmental cost. Replenishment raises property values (a kind of government "givings") and has in a number of cases (e.g., Carolina Beach, North Carolina, and Jacksonville Beach, Florida) led to increased density of development. In such cases, single-family-home communities can find townhouses and high-rise condominiums beginning to hug the shore. It is important for a community to deal with the issue of development density restrictions before the new beach comes in.

Project Promises, Politics, and Predictions

Replenished beaches, by law, must be justified as storm protection for the community. Forming a new recreational beach is not acceptable as the principal justification for emplacement of a replenished beach by the U.S. Army Corps of Engineers. Of course, the reason most communities want replenished beaches is to improve the economy and the quality of life with a wide and handsome new beach, as well as protection for the buildings that profit from the beach traffic.

The Charleston District of the Corps received considerable local criticism about the Folly Beach, South Carolina, replenishment project because the replenished beach disappeared far more quickly than local citizens had been led to believe it would. The Corps had said the beach would not need new sand for eight years (an eight-year nourishment interval). In response to this criticism, district spokespersons claimed that there was nothing to worry about because all the sand was still "in the system," just offshore, still protecting the community from storms—that the beach was working "just as designed." This reply has become a common response by Corps districts to communities where nourished beaches disappear faster than anticipated (e.g., Ocean City, Maryland).

At Folly Beach one year after the beach was pumped in, the Corps said that 95 percent of the beach could still be accounted for. Two years after emplacement, 89 percent was said to be still in place. Meanwhile, in place or not, city administrators and many local residents are unhappy and unimpressed with the underwater sand. The narrow to nonexistent high-tide beach is not what the chamber of commerce hoped for, and the storm berm no longer offers much storm protection. Furthermore, the offshore sand (Corps claims to the contrary) does not offer much storm protection, either.

Figure 5.21 illustrates, in cartoon fashion, what has happened at Folly Beach. Immediately after replenishment, a protective berm exists as well as a wider dry beach (the recre-

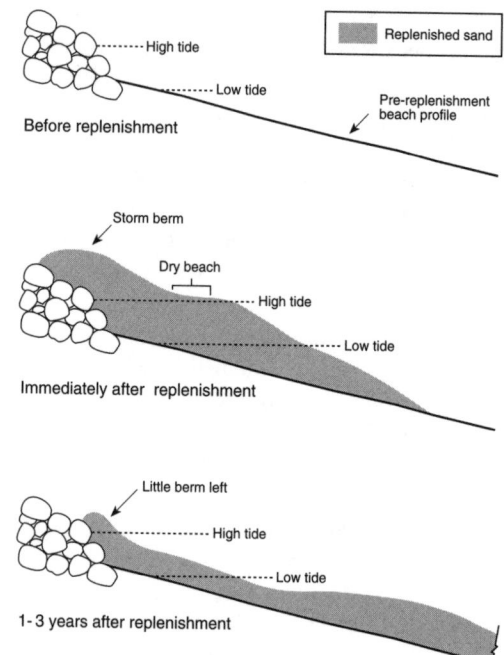

5.21 Cartoon showing a replenished beach before emplacement, immediately after emplacement, and 1–3 years after emplacement. Often the sand from the storm berm is removed offshore, where it plays no role in storm protection for the community. Claims by beach designers that the sand is just offshore, still offering storm protection, are meaningless attempts to justify why a beach was lost more quickly than anticipated.

ational beach at high tide). The profile also is steepened somewhat, and wave energy begins to redistribute the sand out onto the subaqueous portion of the beach. The sand is moved out beyond the surf zone and no longer offers much resistance to storm waves, especially if storm surge occurs, which it almost always does. The protective storm berm is diminished and the high-tide dry beach is very narrow or absent; that is, underwater sand offers no space for playing beach volleyball unless the players can hold their breath for long periods of time. This was the state of Folly Beach approximately two years after completion of the project, with six years to go until the projected "next needed" nourishment.

More replenishment is on the drawing board in South Carolina and other states. Community residents where such projects are scheduled should have a clear understanding of what is in store for them. The Corps' projected nourishment intervals for North Myrtle Beach and Myrtle Beach are 10 years. For Garden City, South Carolina, the Corps says sand will have to be pumped in every six years. These replenished beaches are unlikely to last as long as the Corps has predicted, especially the dry beach and protective berm. Almost certainly the Corps will tell city officials after a year or two that almost all of the sand is in place (offshore), so not to worry. We believe underwater sand doesn't count, and that people should tell this to their local Corps of Engineers district office.

A Test of Beach Replenishment Design Validity

Before entering into a beach replenishment program, a coastal community should be wary of overly optimistic predictions of beach lifespan. Lifespan of a replenished beach is usually expressed as a nourishment interval. For practical purposes, the nourishment interval can be defined as the time when one-third to one-half of the replenished beach above the low-tide line has disappeared. At this point, a new replenishment project is called for.

Values for initial beach nourishment volume, proposed future renourishment interval, and proposed renourishment volume should be readily available in the beach replenishment design documentation. From these data a crude estimate can be obtained of the time the designer expects the beach to survive. If the proposed total lifespan is greater than the lifespan numbers mentioned at the beginning of the section above on beach replenishment, be skeptical. The design is too optimistic and the actual, long-term costs will be many times greater than the proposed costs. For example, replenishment projects currently underway or being proposed in Sea Bright, New Jersey, Myrtle Beach, South Carolina, and Folly Beach, South Carolina, have design lifespans of 30, 36, and 12 years, respectively! These are impossibly optimistic estimates, unless the Atlantic Ocean stops having storms! Less than five years in all cases is more likely!

On the U.S. Gulf of Mexico and Pacific coasts, the best approach to estimate the validity of beach durability predictions is to compare them with the replenishment experience of adjacent shoreline communities. Although a rough method at best, such comparisons are the most accurate approach possible. Remember also that a large storm could wipe out the beach the day after it was emplaced. We will not have accurate long-term predictions of replenished beach lifespans until we can accurately predict the long-term occurrence of storms, which of course means we will *never* accurately predict beach durability, sand volumes required, or costs. Experience is the best teacher. Once a beach has been replenished a few times, the community has a good idea of what the future costs may be and a good basis for judging the feasibility of the replenishment alternatives. A standing rule should be that every replenishment project should include lifetime monitoring as part of the design/plan budget and monitoring costs included in the cost/benefit ratio. Promises will be more accurate, success/failure determined objectively, and a basis provided for adjusting future renourishment plans.

"Free" Beach Replenishment

Sand taken from inlet channel and harbor dredging used to be routinely disposed of by dumping it out on the continental shelf, "out of the way," or piled as dredge spoils on marshes

5.22 A 1994 view of a bulldozer spreading replenishment sand on Atlantic Beach, NC. In the foreground is the fountain formed by the replenishment sand slurry at the end of the dredge pipe.

or on the back sides of islands. Such spoil is a waste of a potentially valuable resource and can be put to good use, as seen in an example from Bogue Banks, North Carolina.

The eastern end of Bogue Banks, including the town of Atlantic Beach, has been the beneficiary of two rather extensive replenishment projects at no cost to the community. Dredge spoil from maintenance of Beaufort Inlet was piled up on the back side of the eastern end of the island. As the spoil dumping ground became full, and a new dredging project was about to begin in 1987, the previously dredged material was taken from the original disposal site and placed on the beach from near the Triple S Pier (the easternmost fishing pier) west to beyond Atlantic Beach (fig. 5.22). The beach

was widened and sand fencing has resulted in modest dune growth. The project was repeated in 1994. The point here is that all dredged sand, if it is at all suitable for a beach, should not be disposed of as spoil but should be used for beach replenishment. Atlantic Beach was very fortunate because their beach was replenished for free. Sand is not likely to be free too often. However, remember the Sand Commandments: If sand has to be dredged for navigation, that sand should be put back into the nearshore sand transport system at another point. Sand disposed of at sea is lost from the beach/dune system.

Free sand may also come with minor drawbacks, as is the case with Atlantic Beach's new beach. The sediment dredged from the inlet channel has a high mud content and contains some oyster shells and shells of other lagoonal bivalves. The shells were protected from abrasion in the muddy, lagoonal environment, so when they are placed on the beach wave action begins to break them up, creating newly broken, sharp, angular shell fragments, potentially dangerous to beach users. Also, the mud in the sediment is cohesive enough to form small scarps on the beach (fig. 5.23) and packs down into a relatively hard pavement on the upper beach—not ideally suited for sunbathing and general beach frolicking. The good news is that the beach is there, protecting development behind it. In addition, as the sediment is reworked, the mud is washed away and the sand content is contributed to the recreational beach.

5.23 Erosion scarp on Atlantic Beach, NC, with oyster shells from the lagoonal sediment source protruding. Such scarps are a common feature on replenished beaches because they erode so quickly. They are a major hazard for nesting turtles.

5.24 Beach scraping on a beach at Nags Head, NC. Such artificial dunes offer temporary protection during storms but are not a long-term solution to the erosion problem.

In recognition that maintenance dredging sand is a valuable replenishment sand resource, the state of Florida has outlawed hopper dredging to prevent loss of sand. The Corps of Engineers continues to dispose of much dredging material offshore, however. Often the dredged sand is not of suitable quality to be used for replenishment, and often there is not a beach needing replenishment nearby.

Beach Scraping (Bulldozing)

Some advocate moving beach sand from the low-tide beach to the upper back-beach (independent of building artificial dunes) as an erosion mitigation technique. A relatively thin layer of sand (1 foot—0.3 meters—or less) is removed from over the entire lower beach using a variety of heavy machinery (drag, grader, bulldozer, front-end loader) and spread over the upper beach (fig. 5.24). The objectives are to:
 —build a wider, higher, high-tide dry beach
  —for recreational use
  —for storm protection
 —fill in any troughlike lows that drain across the beach
 —encourage additional sand to accrete to the lower beach
The newly accreted sand can, in turn, be scraped, leading to a net gain of sand on the manicured beach. The goal of an enhanced recreational beach is achieved for the short term

(e.g., at Topsail Island, North Carolina, and Hilton Head, South Carolina), but the drawback is that *no new sand is added to the system*. Ideally, scraping is intended to encourage onshore transport of sand, but most of the sand "trapped" on the lower beach is brought in by the longshore transport. Removal of this lower-beach sand deprives downdrift beaches of their natural nourishment. Three major negative effects of beach scraping are:
 —interruption of sediment supply (downdrift erosion)
 —steepening of the beach profile
 —complete destruction of beach organisms
The first two of these impacts is known to accelerate beach erosion.

The technique of beach scraping is widely applied in several Atlantic Coast states, but there are few studies of its effectiveness because few projects have been monitored, and those for short intervals only. A general review of scraping projects, including a successful project at Topsail Beach, North Carolina, is provided by Wells and McNinch (1991) and McNinch (1989). They note that beach scraping should be regarded as highly experimental, and any community contemplating using this approach should follow their guidelines:
 —identify erosion problem and recognize limits of scraping
 —assess feasibility (e.g., not feasible for narrow beaches)
 —evaluate fill source/beach slope, cost and design

—include long-term monitoring in project plan/budget

Wells and McNinch (1991) found that scraping was a common emergency response after Hurricane Hugo in South Carolina. As is often the case, under the pressure to repair the beaches quickly, the state's guidelines were not followed and scraping exceeded the recommended 1-foot (0.3 meters) maximum depth. The resulting beaches were of variable configurations.

Protective beach width may be slightly enhanced over the short term, but this gain is insufficient to provide long-term protection against major storms. Scraped beaches may erode faster during storms than adjacent unscraped beaches, although scraping may make a small contribution to post-storm beach recovery (e.g., in Ocean City, Maryland). At best, the old cliché "more study is needed" applies, and beach manicuring describes the technique. Monies put into such programs probably are better spent on encouraging dune growth, bringing new sand into the system, or accelerating design studies for the more substantial beach nourishment projects that will be needed. In Ocean City, Maryland, beach scraping did prove to be a short-term deterrent to beach erosion until the community turned to ongoing beach nourishment, costing over $63 million between 1988 and 1995. Beach sand also is scraped and bulldozed to build dunes at the back of the beach.

Dune Building

Coastal dunes are a common landform at the back of the beach, part of the dynamic equilibrium of barrier island beach systems. Although an extensive literature exists for dunes (e.g., Nordstrom, Psuty, and Carter, 1990; Psuty, 1988), their protective role often is unknown or misunderstood. Frontal dunes are the last line of defense against ocean storm wave attack and flooding from overwash, but interior dunes may provide high ground and protection against penetration of overwash and against the damaging effects of storm-surge ebb scour. A detailed study of the South Carolina coast after Hurricane Hugo (Thieler and Young, 1991) quantified the coastal geomorphic changes induced by the storm and demonstrated the protective role of dunes (tables 5.4 and 5.5). That study concluded that the minimum dune field that survived Hurricane Hugo, and thus protected buildings, was 100 feet (30 meters) wide with dunes about 10 feet (3 meters) high. Most of the buildings damaged or destroyed by Hurricane Hugo were fronted by beaches less than 10 feet (3 meters) wide and dune fields less than 50 feet (15 meters) wide. An important ramification of the Thieler and Young (1991) study is that it provides an objective basis by which to predict damage in other coastal areas for a similar storm.

Prior to the passage of Hugo, 70 percent of the 51-kilometer stretch (about 30 miles) of developed South Carolina shoreline impacted by the storm was classified as dune field, that is, one or more continuous, well-vegetated dune ridges. After Hugo, only 15 percent (less than 8 kilometers—about 5 miles) of the study areas fell into this category (see table 5.4). Hugo destroyed 17 miles (over 28 kilometers) of dune fields in just a few hours!

Wide dune fields (both natural and bulldozed) were a main line of frontal defense against property damage during Hugo. Dune field widths were narrowed by Hugo (see table 5.5), but where the widest and highest dunes existed, property damage was less. A price was paid, however. Almost half of the study area shoreline (49 percent) had dune fields greater than 100 feet (30 meters) in width before Hugo. After Hugo, only 21 percent of the study area had dune fields greater than 100 feet wide remaining.

Beach width also played a role along with dune field width and height in protecting property during Hurricane Hugo. The Thieler and Young (1991) study showed this relationship for the areas classified as "new gap" after Hugo. New gaps are defined as areas where buildings were completely destroyed and/or removed from their foundations by the storm. The vast majority (84 percent) of buildings so damaged were fronted by the "deadly" combination of beaches less than 10 feet (3 meters) wide and dune fields less than 50 feet (15 meters) wide. Recall that beach widths are consistently less in front of seawalls.

The conclusions based on a specific coast for a specific storm, in this case Hurricane Hugo

Table 5.4   South Carolina Shoreline Before and After Hurricane Hugo

	Shoreline Length (km)	Before (%)						After (%)					
		df	db	b	wt	bh	r	df	db	b	ws	wf	ng
Garden City	7.8	24	23	5	0	42	6	0	0	0	100	0	14
Litchfield	6.6	53	27	0	16	4	0	12	8	1	71	8	0
Pawleys Island	5.9	9	68	1	7	15	0	9	12	14	65	0	12
Debidue Beach	6.0	31	47	0	0	22	0	60	0	0	38	2	2
Isle of Palms	10.0	90	1	0	0	0	0	0	0	0	100	0	4
Sullivans Island	4.8	94	0	0	0	1	5	0	0	0	65	35	4
Folly Beach	9.9	18	20	1	0	2	59	14	0	0	85	1	11
Total:	51.0	45	25	1	3	12	14	13	2	2	78	5	7

df = dune field, db = bulldozed dune, b = no dune/beach only, wt = washover terrace, bh = bulkhead, r = revetment, ws = washover sheet, wf = washover fan, ng = new gap.
Source: Thieler and Young, 1991.

Table 5.5   Width of Frontal Dune Before and After Hurricane Hugo (South Carolina)

	Shoreline Length (km)	Before (%)				After (%)			
		A	B	C	D	A	B	C	D
Garden City	7.8	45	25	13	17	100	0	0	0
Litchfield	6.6	4	31	16	49	79	21	0	0
Pawleys Island	5.9	24	25	20	31	78	6	3	13
Debidue Beach	6.0	0	12	4	84	14	7	2	77
Isle of Palms	10.0	0	5	17	78	33	15	23	29
Sullivans Island	4.8	4	6	1	89	19	30	8	43
Folly Beach	9.9	34	37	14	15	71	15	7	7
Total:	51.0	17	21	13	49	59	13	7	21

A = < 1 m, B = 1–15 m, C = 15–30 m, D = > 30 m.
Source: Thieler and Young, 1991.

in South Carolina, may not necessarily be fully applicable to other geologic settings and storms of different intensities, tracks, and durations. However, Hugo provided a general experience for predicting property damage in other developed areas. Obvious high-damage areas will most certainly be shoreline stretches with narrow beaches, low elevations, and narrow or absent dune fields.

Dunes are critical coastal geomorphic features with respect to property damage mitigation. Prior to strict coastal zone management regulations, frontal dunes were often excavated for ocean views or building sites or notched at road terminals for beach access. These artificially created dune gaps are exploited by waves and storm surge (fig. 5.25) and by storm-surge ebb flows. Wherever dune removal for development has occurred, the probability is increased for the complete overwash and possibly inlet formation. The combined threats of storm flooding, inlet formation, and the burial of roads by overwash sand make areas of dune removal prime danger zones for evacuation in case of a hurricane warning. Dune gaps can be refilled (plugged), many with just a few truckloads of new sand. As we will see in chapter 6, maintaining dunes in the interior of islands is an equally important means of property damage mitigation.

Plugging Dune Gaps

Notches cut in dunes for beach access, views, and housing sites are naturally exploited by waves and storm surge, increasing the potential for storm damage to development behind the notch. Plugging dune gaps is a simple and relatively inexpensive measure to mitigate the effects of these artificially created overwash passes.

Nags Head, North Carolina, has several stretches of shoreline where dunes are missing, mostly in front of motels. Although the gaps are large, they could be plugged to increase protection from moderate storms. Another large gap in the very center of Nags Head has possible implications for evacuation as well as for property damage. The only mainland access (and thus, evacuation route) to Nags Head and Hatteras Island to the south is U.S. Route 64, which stretches across Roanoke Sound and the island of Manteo before reaching the island at Whalebone Junction. The portion of North Carolina's Outer Banks from the Virginia border to Oregon Inlet, a distance of over 55 miles (90 kilometers), is not technically an island, but a spit. Numerous historical inlet locations, however, are testimony to the fact that this was a true island in the past and will likely be again in the future (see chapter 7). The processes are the same as for a true barrier island, and we use the terms "island" and "spit" interchangeably here. This low-elevation area was made more vulnerable because essentially

5.25 A fresh overwash fan that has just been added to the land by waves flowing through a dune scarp near Islote, Puerto Rico (north coast).

5.26 Whalebone Junction in Nags Head, NC, where a large dune gap exists at the landward end of the pier that could furnish a path for storm waters to cut off the highway escape route from the island. This gap could easily be repaired with a few dump trucks of sand. Photo by Rob Young.

all of the dunes were removed for development (fig. 5.26). The potential flooding of Whalebone Junction several hours before a storm hits, and while evacuation will still be ongoing, is exacerbated by the lack of dunes. Rebuilding the dunes to close the large gap here is a very inexpensive way to "buy" several more hours of safe evacuation time when the inevitable storm bears down on Nags Head.

Sand fencing and artificial plantings of dune grass to build (or rebuild) dunes is most effective when property is set back far enough to provide adequate space for dunes to build and stabilize in a natural equilibrium profile and location. In a case like Bogue Banks it helps that the island has a very high sand supply (meaning that geologic conditions are right for a lot of sand to be moving onshore). In contrast, islands with low sand supplies defy efforts to build dunes artificially. The post–Hurricane Hugo sand fencing project at Folly Beach, South Carolina, trapped little sand and much of the fencing quickly washed away.

Dune Management

Having said that dunes provide natural protection and that dunes can be repaired or restored artificially, some qualifications must be stated. A fundamental rule is that "the dune zone may have to migrate if it is to retain its protective characteristics" (Psuty, 1988). Unfortunately, some people view dunes as if these natural sand accumulations can be designed, engineered,

5.27 Post–Hurricane Hugo sand fencing on Garden City, SC. The fencing pattern used here is a zigzag, but the most efficient fencing pattern may vary from location to location.

and constructed in the same fashion as groins and seawalls. If dunes are to function in their natural protective role, the PAR approach must mimic the equilibrium setting. A dune bulldozed into a nonequilibrium location will not be stable (see fig. 5.24). In fact, artificially constructed sand dikes (continuous dune lines as constructed on the Outer Banks during the Great Depression), rock-cored "dunes," and dunes artificially "cemented" in place are like seawalls blocking sediment derivation and migration, creating new problems on other parts of the island, including the interior. Dunes are not static, rigid, permanent structures.

Similarly, the natural vegetation that stabilizes dunes is in an equilibrium with the sedi-

ment and biota. The roots of sea oats (*Uniola paniculata* L.) are colonized by vesicular-arbuscular mycorrhizal (VAM) fungi, which improve the grass's uptake of nutrients and water (Sylvia, 1989). Artificial plantings of these grasses do not fare well in pumped-in nourishment sands or where these sands have been used to construct artificial dunes because the VAM fungus is missing. Plant root systems can be inoculated with the VAM fungus in the nursery to improve colonization of the artificial plantings (Sylvia, 1989), but the point is that nature is far more complex than a simple engineering design. Even the differences in grain size and sorting between bulldozed sediment and natural wind-blown sand may influence dune stability and plant growth.

For additional guidelines to building and vegetating dunes see Seltz (1976), Woodhouse (1978), Broome, Woodhouse, and Seneca (1982b), SCMEP (1990), and Texas General Land Office (1991).

Principles of Sand Fencing

Dune gaps and dune lines can be augmented and repaired by encouraging or enhancing natural processes (fig. 5.27). Typically, such augmentation includes the placement of sand fences, recycled Christmas trees, and the planting of dune grasses to trap wind-blown sand. Sand fencing also serves as a barrier to foot traffic over dunes, allowing vegetation to gain a foothold and flourish. To be effective in

building dunes, the use of fencing should follow guidelines that have grown out of various studies (e.g., Broome, Woodhouse, and Seneca, 1992b; Psuty, 1988) and local experience:
(1) Do not fence on beach:
  —landward transport of sand may be reduced
  —the small dunes buildup is out of equilibrium
  —fencing is likely to wash out in the first storm
(2) Fencing on the face of the dune may encourage seaward dune toe growth.
(3) Do not use double rows of dune fencing which act as inpenetrable rather than permeable barriers (blocks inland sand flow).
(4) Do not attach new fence to old fence as dune grows upward.
(5) Remove fencing periodically.
(6) Vegetate the newly formed dunes with native species.
(7) Vegetation is just as effective as sand fences in building dunes over the long term.
(8) Trapping sand on the landward side of the dune is as important as trapping sand on the front and top of the dune. This last guideline is very important because dune width, as well as dune height, affords protection. The principle is often violated when property owners notch the back side of the dune (e.g., for driveways, patios, play areas). Post-storm reconstruction is a time when the landward sides of dunes

suffer loss of sand and reduction of width. Instead of placing overwash sand from roads, driveways, and parking lots back onto the dune, the sediment is removed. Remember:

—both dune width and height are important
—keep all the sand in the system (add new sand)
—allow the dune to migrate (maintain equilibrium)
—vegetate with native species

Unneeded Dune Building

A sand fencing project completed after Hurricane Hugo along Sunset Beach, North Carolina, consisted of a continuous row of sand fencing and was emplaced so far seaward on the beach that there is almost no dry recreational beach at spring high tides (fig. 5.28). Sand fencing was probably not needed on this beach because it has a history of natural accretion. In the present situation, the fencing is trapping sand too far out on the beach rather than where the sand would have accumulated naturally, just a few tens of meters (100 feet or so) landward at the edge of the maritime shrub vegetation. The current ridge of accumulating sand will be vulnerable to wave attack in future minor storms. The sand fencing should have been placed properly at least 15 feet

5.28 (a) Sand fencing at Sunset Beach, NC, six months after Hurricane Hugo. This straight fence has caused sand to accumulate behind it but also prevented turtles from nesting. (b) The same area one year after Hugo. Christmas trees have been added to the fence and the grasses have obviously flourished. It is questionable whether Christmas trees are really a good option for accumulating sand, in part because dead trees, especially ones with tinsel, detract greatly from the aesthetics of a beach.

(about 5 meters) from the spring high-tide, winter storm swash line. Sand fences are neither necessary nor desirable in all situations.

After Hurricane Hugo, an emergency "dune" (really just a sand pile) was emplaced along much of the impacted portion of the South Carolina shoreline, theoretically to protect structures from additional damage due to oncoming extraordinarily high (but still "normal") astronomical tides. The total cost for this emergency operation was $15 million. Federal disaster relief funds were used to pay some of the cost. Remnants of those dunes still exist because they were subsequently vegetated. How much "protection" these dunes afforded is uncertain because they were only ridges of unconsolidated sand, not stabilized by vegetation, at the time of the high tide. It is not clear if the cost of emplacing the sand ridges was worth the small level of protection they could have afforded.

Living with an Accreting Beach

Sunset Beach, North Carolina, has a problem all oceanfront communities would like to share. This example is included because of the remarkable history of accretion. The island is of low elevation and narrow with high-density single-family-home development (similar to many barrier islands), but rapid accretion is occurring along the central part of the island. The maximum long-term average annual shoreline change as calculated by the North

Carolina Division of Coastal Management is over 8 feet (2.4 meters) of accretion per year in the central portion of the island. This average erosion/accretion rate is determined over a period of 50 years—the length of time that aerial photographs suitable for documenting shoreline changes have been available. Accretion rates tail off on either side of the central part of the island, and the island's ends have actually undergone long-term erosion.

The beach has built out so much in the vicinity of the fishing pier that the pier had to be extended. Accretion has led to the development of a wide dune field, though the dunes are of relatively low elevation. The original dune field consisted of four poorly defined rows of dunes or beach ridges. Hurricane Hugo removed two of these beach ridges. These dunes were sacrificed to the storm, but they dissipated wave energy, sparing the houses behind them. The post-Hugo situation was one of a flat beach, a typical post-storm beach profile. Accretion began again as before Hugo and post-storm sand fencing was emplaced to help build up the new dunes (see fig. 5.28). It may take years, however, for an appreciable volume of sand to accumulate and for the dunes to reach their pre-Hugo size. Current accretion or seaward buildup is generally unusual on barrier islands anywhere in the world, other than spits or islands associated with major river deltas where the effects of high sediment supply are greater than those of the sea-level rise and wave regime. When conditions change, accretion will

be replaced by erosion on Sunset Beach in the coming decades.

No one is exactly sure why Sunset Beach is accreting or if it will continue. Most likely, the accretion is at least partially influenced by tidal inlet dynamics. Sunset is a very short island, and most of it is protected at least somewhat by ebb tidal shoals that block some of the incoming wave energy. If inlet dynamics change, Sunset Beach may change to an erosive mode. Another possibility is that the artificial relocation of Tubbs Inlet resulted in a new local sediment source and/or change in wave refraction, leading to the accretion.

Similar areas of accretion occur on the updrift sides of major jetty systems (e.g., Sullivans Island, South Carolina, and Ocean City, Maryland). Where a jetty blocks the longshore drift, beaches widen and dune ridges may accumulate.

Soft Stabilization: A Final Word

Soft approaches to dealing with shoreline erosion take many forms, ranging from simple dune grass planting to massive beach replenishment projects. While generally considered a more environmentally sensitive approach to erosion management, questions of long-term impact still remain (see table 5.6). In addition, as suitable sources of replenishment sand become more scarce and more distantly located, the costs and environmental impacts increase.

Dune protection and improvement guidelines are being developed by several states and private consultants. Keeping people off the dunes with prohibitions, dune walkovers, and barriers is the important first step in dune conservation. Many states have strict rules governing how to build, vegetate, and maintain artificial dunes. Guides to building and vegetating dunes are numerous (e.g., SCMEP, 1990; SCCC, undated).

Beach replenishment is being billed as the wave of the future for shoreline protection and coastal management. We hope we have shown that the costs, in dollars and in environmental concern, may outweigh the benefits. Like most things, if it sounds too good to be true, it probably is.

Relocation: Move 'Em off the Line

The most obvious way to avoid a hazard is to stay away or move away from it! So it is with an eroding or shifting shoreline. The prudent planner, recognizing that over the years erosion is likely to occur, will build well back from the shoreline or shore bluff. How far back is a "safe" building setback? That question is difficult to answer and will vary from place to place according to erosion rates and state and local regulations.

While building setback puts some distance between your property and the shore, you cannot expect that distance to remain constant. Shorelines are not fixed entities, but are migrating. When the shoreline "catches up" to

Table 5.6 Advantages and Disadvantages of Beach Replenishment

Advantages
The beach is widened
Temporary protection of property
Storm protection
Maintains the recreational value of the beach
No negative impact on downdrift beaches
 (becomes a better sand source)
Looks better (aesthetics)

Disadvantages
Temporary, must be renourished
High cost (increases with each new nourishment)
Decreasing sand supply to maintain project
Possible damage to marine organisms by turbidity
Offshore dredge hole may create erosive wave
 refraction pattern

your property, the original setback distance won't do you any good. That's when relocation of threatened structures makes sense.

The most obvious type of building relocation is actually picking up a building and moving it somewhere else, either in one piece or in sections. However, relocation of buildings can also mean demolishing and rebuilding somewhere else. Any active or passive method of moving, or abandoning and rebuilding, is essentially what is meant by building relocation (see table 5.1). Abandonment may be an economically sound option, especially when the building has existed well beyond its design life and where the cost of moving, or protecting

the building in place, exceeds the building's value. The obvious advantages of relocation are that the beach is preserved, buildings are preserved, and shoreline stabilization costs are saved for the property owner, the community, and the taxpayers. The major drawbacks are that relocation can be politically difficult as well as costly, and that land is ultimately lost (Pilkey, 1991). Relocation has been employed on the coast for almost 150 years. With time, it will become more and more accepted, particularly as the preservation of beaches becomes a more accepted and important aspect of coastal management.

When a building is threatened by erosion, the costs and benefits of moving the structure back from the shore must be weighed along with other alternatives such as hard or soft stabilization. Depending on the nature of the problem, a move-back can compare favorably to these alternatives and prove to be economically and aesthetically superior in the long run.

An amendment to the National Flood Insurance Program (NFIP) passed in 1987 (the Upton-Jones Amendment) allowed homeowners of threatened buildings to use up to 40 percent of the federally insured value of their homes for building-relocation purposes. The law recognized relocation as a more economical, more permanent, and more realistic way of dealing with long-term erosion problems. The federal government (taxpayers at large) would pay a relatively small amount to assist in relocating a threatened house rather than paying a larger amount to help rebuild it, only to see the rebuilt house destroyed in a subsequent storm, and paying to rebuild again . . . and again. By March, 1995, North Carolina had claims for over 70 relocations and 168 demolitions, and accounted for over 60 percent of all coastal claims under the program.

The National Flood Insurance Reform Act of 1994 terminated the Relocation Assistance Program as of September 23, 1995, and replaced the Upton-Jones program with the National Flood Mitigation Fund. Financed from penalty revenues collected for noncompliance with NFIP requirements, the new program provides state and local governments with grants for planning and mitigation assistance for activities that will reduce the risk of flood damage to structures covered under the NFIP. Demolition and relocation activities are eligible for grant assistance under the program, but now compete with other mitigation approaches, including elevation and flood-proofing programs, acquisition of flood zone properties for public use, beach nourishment activities, and technical assistance. Limits are placed on how much a state or community can receive in a five-year period under the program (e.g., $10 million for a state; $3.3 million for a local community; $20 million collectively within any one state).

If these monies continue to go to relocation programs, or activities that permanently lower risk, the dollars will be well spent. If they are siphoned off for beach nourishment projects that only delay the inevitable loss, no progress in mitigation will have been made. And, of course, the questions can be raised: What's the government doing in the insurance business? Why should the taxpayers be bailing out people who choose to invest in very expensive, and very risky, property? The answers include trying to get property owners to pay part of the disaster cost before the fact, rather than taxpayers' being expected to cover all of the cost after the fact; trying to save the taxpayers some money and educate purchasers; trying to save us from ourselves.

Remember that the cost of moving back is likely to be a one-time expense, whereas hard and soft stabilization approaches will be continual expenditures, plus the cost of ongoing maintenance. Furthermore, the big storm most likely will inflict damage to your property in spite of these precautions. In the case of some structures, letting the sea claim the building when its time comes may be the most realistic solution, both economically and environmentally. Relocation is a recommended means of shorefront management only if economically viable. To date, mostly single-family houses and small commercial structures have been relocated. The cost to relocate multistory buildings may be prohibitive in many settings, another reason why communities should exert control over types and sizes of buildings. Another approach is to demolish threatened structures and rebuild them elsewhere.

5.29 The beautiful Cape Hatteras Lighthouse exists only because of three groins (one of which is shown here) and the sandbag revetment.

Cape Hatteras Lighthouse: Move It or Lose It!

The controversy surrounding the options for preserving the Cape Hatteras Lighthouse presents a microcosm of shoreline management issues, particularly with respect to relocation. Located in the Cape Hatteras National Seashore (see fig. 1.2), the lighthouse is owned by the National Park Service. Arguably the world's most famous lighthouse, the Cape Hatteras Lighthouse stands 208 feet tall, the tallest brick lighthouse in the United States, and weighs 2,800 tons (fig. 5.29). The light has helped warn mariners of the treacherous waters that have given North Carolina's Outer Banks the nickname "Graveyard of the Atlantic."

Table 5.7   The Saga of the Cape Hatteras Lighthouse

1870	Existing lighthouse first lighted; at 208 feet (69 meters) it is the tallest brick lighthouse in U.S. Original distance from the sea: 1,500 feet (450 meters).
1919	Shoreline within 300 feet (90 meters) of lighthouse.
1935	Shoreline migration brings the sea to within 100 feet (30 meters).
1936	Coast Guard abandons lighthouse. Light moved to steel skeleton tower in Buxton Woods, 1 mile west. Erosion control attempted with construction of sheet-steel piling.
Late 1930s	Civilian Conservation Corps begins dune-building project to prevent overwash and allow future development behind it.
1950	Shoreline stabilized (naturally and temporarily) and Cape Hatteras Lighthouse reactivated by Coast Guard. Ownership of the structure transferred to the National Park Service (NPS).
1966	312,000 cubic yards of sand pumped from Pamlico Sound to stabilize the shoreline.
1967	Nylon sand-filled bags emplaced in front of lighthouse to stabilize the shoreline. (Some still remain in 1995.)
1969	U.S. Navy builds three groins to protect naval facility and lighthouse. They are destroyed by storms and rebuilt in 1975.
1971– 1973	Two replenishment projects emplace 1.5 million cubic yards of sand from Cape Hatteras Point to the lighthouse area. September 1973 finds the sea 175 feet (50 meters) from the old lighthouse ruins and 600 feet (180 meters) south of the present lighthouse.
1978	Water reaches old lighthouse ruins.
1980	March storm washes away remaining ruins of the original lighthouse and water reaches within 70 feet (20 meters) of present lighthouse.
1980	During the summer, NPS receives results of study of Cape Hatteras erosion problem and asks the U.S. Army Corps of Engineers to begin evaluation for methods to preserve the light.
1982	Public workshop held April 1–2 in Manteo, NC, to discuss alternatives for protecting the lighthouse. Options include a seawall revetment, offshore breakwaters, beach nourishment, additional groin, relocation, and no action.
1985	NPS selects seawall revetment as best option.
1986	Move the Lighthouse Committee organizes.
1987	NPS decides to review options, asks the National Research Council (NRC) for help.
1988	NRC final report unanimously selects relocation as the best option.
1989	In early summer NPS promotes relocation as the preferred alternative and again asks for public input. In December NPS formally announces their decison that relocation of the lighthouse is the best way to preserve it.
1991	NPS seeks bids for moving designs.
1994	Construction of a fourth groin to protect the southern exposure of the base of the lighthouse begins.
1995	Still waiting to be moved. . . .

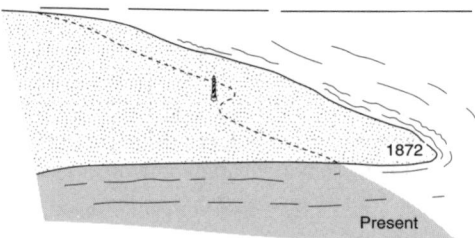

A. Comparison of 1872 and present shorelines

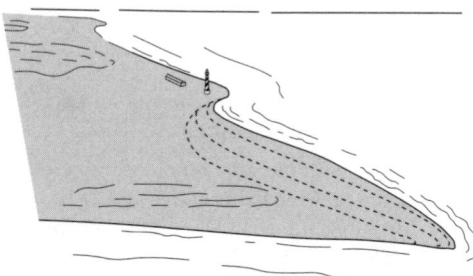

B. What will happen if lighthouse stays in place

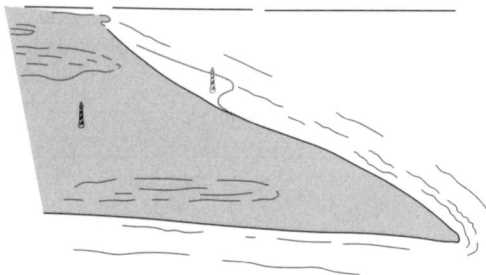

C. What will happen if lighthouse is moved

The present light at Cape Hatteras was first lighted in 1870 (table 5.7); it replaced a smaller lighthouse that had far less illuminating power. Since the 1930s, when the present light was first seriously threatened by shoreline erosion, until 1981, the National Park Service (NPS) spent about $15 million on interim protection methods. Many of these shoreline projects were primarily for protection of a U.S. Navy facility located just to the north of the lighthouse and included groins, beach nourishment, and sandbagging. The shape of the shoreline around the lighthouse bulges unnaturally because of the groin field. In 1980 when the light was almost lost to a winter storm, NPS began investigating methods of "long-term" protection in order to find a "solution" to the erosion problem.

NPS was directed by the Department of the Interior to find a protection method that would meet three criteria: (1) the lighthouse would be saved; (2) the solution would be permanent; and (3) there would not be major recurring costs. Despite all the controversy, an examination of all facts clearly showed that

5.30 The saga of the Cape Hatteras Lighthouse. (a) The lighthouse was originally 1,500 feet (450 meters) from shore. (b) Groins trap sand, protect structures to north, create erosion to south. (c) New location of lighthouse will reestablish its original relationship to shoreline. Deterioration of groins will allow reformation of natural shoreline, causing motels to fall in.

only moving the lighthouse satisfied all three criteria. That conclusion was reached by the Move the Lighthouse Committee which, in 1987, helped convince NPS to reexamine the issue. The same conclusion was also reached by the Committee on Options for Preserving the Cape Hatteras Lighthouse, formed by the National Research Council in July 1987 at the request of NPS. The chronology taken from NPS's Environmental Assessment for the Lighthouse Protection Plan of 1982 is presented in table 5.7.

Figure 5.30 compares the 1872 and present-day shorelines, depicting over 1,500 feet (450 meters) of shoreline retreat, and demonstrates that if the lighthouse is not moved, the shoreline will continue to erode past the lighthouse and the curvature in the shoreline south of the lighthouse will become even more pronounced. Costs to maintain the shoreline at the lighthouse and to the north will continue to increase. Eventually, the lighthouse will be destroyed in a storm and all the money and effort spent to stabilize the shore will have been wasted. In contrast, if the lighthouse is moved, it could be placed in the same position relative to the shoreline as when first constructed and will no longer need to have the shoreline stabilized. The groins and sandbags will be removed or destroyed by storms, at which time the shoreline will straighten and quickly assume its normal equilibrium profile and shape. This adjustment in shoreline shape will incorrectly be called erosion by some. All other

5.31 A 1925 aerial view of Cape Henlopen, DE. The lighthouse fell into the sea one year later, 170 years after it was constructed. When first built, the lighthouse was about 1,400 feet (420 meters) from the shoreline. Photo by William H. Hoyt.

structures threatened by shoreline adjustment and migration can and should be moved along with the lighthouse or later, as they become threatened.

The main point made by this example is that we as a society cannot afford to stand and fight the sea on all of our coasts (there are dozens of lighthouses to be "saved"). We must plan an organized retreat from the encroaching sea or alternatively face expending vast amounts of money and other resources only to fail and re-

treat grudgingly in a disorganized fashion. Moving the Cape Hatteras Lighthouse will set a bold example for all coastal zone managers to follow. If a lighthouse can be moved, then most buildings can be moved (technically speaking). Some would argue that no public money should be spent on the structure, that it should go the way of the Morris Island, South Carolina, lighthouse (see fig. 4.16) or the Cape Henlopen, Delaware, lighthouse (fig. 5.31). Similarly, some argue that taxpayers in Dallas or Des Moines should not have to pay to protect private buildings in Ocean City or Shoreville.

The 10/100-Year Relocation Concept

The crux of the problem exemplified in areas such as the Myrtle Beach Grand Strand, South Carolina, Miami Beach, Florida, and other great oceanfront resort communities is the vast number of high-rise condominiums and hotels right on the shoreline (see, e.g., figs. 5.16 and 5.20). For the present, beach replenishment is economically feasible for these communities because of the large number of people that use the beaches and the enormous amount of revenue generated. The Miami Beach replenishment project, the most successful on the East Coast in terms of replenished beach lifetime, has lasted for over 10 years (Pilkey and Neal, 1988). In the Grand Strand, replenishment has to be repeated almost yearly. There will come a time, however, when the economics of replenishment will no longer be acceptable. As time goes by, more and more sand will be needed for each replenishment project. The cost per project will continue to escalate. The time is approaching when serious consideration will have to be given to relocation.

Does it sound farfetched to move large buildings? The International Association of Structural Movers says that moving large structures is technologically feasible, though expensive. Recall also that relocation can mean demolishing the building and rebuilding it elsewhere. The unanswered question is an economic one.

Owners of large buildings should begin researching the economics of the various options. One possibility we term the *10/100-year relocation plan*, in which a relocation strategy is developed within ten years and implemented as necessary over the next century. Background work should be done as quickly as possible, certainly within the next ten years. If the replenishment option is to be continued, financing requirements and nonlocal sand sources need to be identified, the sand resources acquired, and a timetable for obtaining all the necessary permits established. Cost comparisons of traditional relocation or relocation by demolition and rebuilding must be considered, especially in light of the temporary nature of beach nourishment and finitude of sand supplies. It must be ascertained if buildings can be relocated on the present property or off property, within the community or out-

side. What are the options and questions yet to be raised? The point is that planning must begin now, so that the proper questions can be addressed. All this groundwork must be completed within ten years so that the plan can be implemented when needed. Such implementation will vary with each island and community but will certainly be needed for all within 100 years (hence the plan's name). Remember, this problem is being passed on to the next generation, who, in turn, will pass it on to their children.

The Grand Strand and Miami Beach are examples of heavily developed beach communities that will face similar problems in the coming decades. Whatever is done, management actions must be in keeping with regional conditions. Coastal cities and communities must work together in the larger context of natural systems. Sand taken from an inlet and placed on one community's beach is sand that might naturally have traveled down the coast to another community's beach. Long-term effects of the mitigation options are not known, but it is known that taking sand from one stretch of shoreline and placing it on another is nothing more than robbing Peter to pay Paul: gaining beach at the expense of one's neighbor. All sand used for replenishment must come from nonlocal sources, preferably from the mainland, or well back from the beach/dune system in the case of the Grand Strand.

The Need for Long-Term Relocation Planning: Bogue Banks, North Carolina

A situation to compare to Myrtle Beach, South Carolina, is illustrated in the community of Indian Beach on Bogue Banks, North Carolina. The Summerwinds Condominium complex sits on the ocean side of Indian Beach. Over 200 units are located here and the seaward-most buildings sit less than 50 feet (15 meters) from a dune scarp. The shoreline is retreating at an average of about 2 feet (0.6 meters) per year. However, the shoreline tends to go back in jumps; for example, the shoreline retreated some 20 feet (6 meters) during a winter storm in 1987.

The situation here is different from Myrtle Beach because there is only one lone condominium complex sitting next to trailer parks and single-family houses. The density of development is not as great as at Myrtle Beach, nor is the value of the real estate as high. The 10/100-year relocation plan also is recommended for Summerwinds and other large condominiums and hotels on Bogue Banks, though the economics will certainly work out differently than for Myrtle Beach.

Nags Head, North Carolina: Town on the Move

The retreat philosophy has been successfully implemented by the town of Nags Head, North Carolina, located about 50 miles (80 kilometers) north of the Cape Hatteras Lighthouse (see fig. 1.2). This mitigation strategy stems from a desire to protect Nags Head's family beach atmosphere which attracted the residents in the first place, according to town planner Bruce Bortz. The town adopted building standards more restrictive than required by either FEMA or the North Carolina Coastal Area Management Act. Incentives are used to encourage development to be located as far back from the ocean as possible. Because small, single-family structures are much easier to move, the town has limited the development of oceanfront hotels and condominiums. Deep lots running perpendicular to the shore provide considerable room for relocation. The general theme of Nags Head's mitigation plan is the recognition that shoreline retreat is inevitable and that it is far better to adopt a policy of planned retreat than to wait for a disaster to force retreat.

The costs of the two "soft" solutions legal in North Carolina, relocation and beach replenishment, were compared by Williams (1993). In Nags Head, the area of highest erosion is from Whalebone Junction to the town's southern border (NCDCM, 1992), a distance of about 4.5 miles (7.2 kilometers). Replenishment studies carried out by the Program for the Study of Developed Shorelines at Duke University show that for a relatively high wave energy area such as Nags Head, the cost of beach nourishment will be approximately $2 million per mile with additional nourishment required every three years—a current average annual

5.32 The John Yancey Motel on Bogue Banks, NC, was built while the memory of Hurricane Hazel (1954) was still fresh. At the time of construction, three dune ridges separated the motel from the sea. At the present time, two dune ridges still exist in front of the motel.

cost of $3 million. This expenditure will continue as long as replenishment is the chosen mitigation technique. Because much of the nourishment sand is likely to come from Oregon Inlet to the south, the response of Oregon Inlet to the currently proposed jetties will have a considerable impact on future costs. Thus, not only are the present costs of nourishment high, future costs are certain to increase.

By comparison, the cost of removing structures from the threatened areas is much lower. As of the early 1990s, Nags Head had accounted for 78 of the 379 (21 percent) Upton-Jones petitions submitted nationwide, 55 of which had been approved (Williams, 1993). Of these 55, 35 requested funds for demoli-

tion and 19 requested funds for relocation. Average costs were $74,409 per structure demolished and $30,211 per structure relocated (Williams, 1993).

Removal costs to date are dramatically less than the cost of beach nourishment for the 4.5 miles (7.2 kilometers) of South Nags Head. Furthermore, beach nourishment will need to be repeated every three years. In contrast, if all the threatened structures are removed, it will be 20 to 25 years before the number of threatened structures returns to current levels (according to Nags Head planners' predictions based on N.C. Division of Coastal Management average annual erosion rate data). Thus, the cost difference between the two plans is dramatic. Beach nourishment would cost about $9 million every three years and the retreat option would cost about $2 million every 20 to 25 years.

Setback for Protection

On Bogue Banks, North Carolina, the Pine Knoll Shores area includes some of the highest, widest, and most stable areas of the island. Much of this area is suitable for development so long as the natural environment is maintained, especially the dense maritime forest in the elevated, middle to back areas of the island. Although the frontal dune is high and continuous, it is narrow and eroding, therefore, the beachfront is not suitable for development. Buildings on the ocean side of the island

5.33 Aerial photo of a portion of the communities of Indian Beach and Salter Path on Bogue Banks, NC, showing development hugging the beach (to the right) in contrast to other development set well back from the eroding shoreline. The latter would be the best property to purchase for those concerned with safety of life, limb, and property.

should be set back from the frontal dune and troughlike depressions behind parts of the dune line. A good example of this is the John Yancey Motel. This motel was built shortly after Hurricane Hazel and was located well back from the water (fig. 5.32). The builders apparently learned a lesson from Hazel and have respected the integrity of the natural topography and vegetative cover.

Deep Property Lots for Future Relocation

Some of the deepest property lots anywhere on Bogue Banks are located in Salter Path, including lots greater than 1,000 feet (90 meters) in depth off Salter Path Road (Route 58) (fig. 5.33). Deep lots allow homeowners to relocate houses threatened by erosion to another location on their own property. In effect, lot depth determines possible future on-site relocation.

Relatively deep lots are found on some other islands (e.g., Pawleys Island, South Carolina), but the norm is to crowd as many rows of homes as possible near the water. Despite this trend, some communities, such as Nags Head, North Carolina, are now requiring people to purchase ocean-to-lagoon lots in order to provide for relocation. Such forward thinking prepares for the inevitable.

Relocation: A Final Word

The 1985 Skidaway Conference on America's Eroding Shoreline (Howard, Kaufman, and Pilkey, 1985) brought together scientists, engineers, attorneys, planners, and environmentalists to discuss a national strategy for beach preservation. As the participants of that conference so eloquently stated over a decade ago:

> Sea level is rising and the American shoreline is retreating. We face economic and environmental realities that leave us two choices: (1) plan a strategic retreat now, or (2) undertake a vastly expensive program of armoring the coastline and, as required, retreating through a series of unpredictable disasters.

Relocation is a viable coastal management tool and need not be considered only for single-family houses. When you move the structure, the danger is reduced (table 5.8).

System-Oriented Mitigation: Preserve, Augment, and Restore Nature

Although the traditional focus has been on the shoreline and island front, the area behind the shoreline is where we live, where we invest our time and financial resources, and where we are in the path of the dynamic processes of the barrier island. Management and mitigation

Table 5.8   Advantages and Disadvantages of Relocating Buildings Back from a Retreating Shoreline

Advantages
　Removes threat to building
　Allows natural shoreline processes to continue
　Preserves the beach
　Good possibility of one-time-only cost

Disadvantages
　High cost
　Site must be deep enough to allow suitable moveback, or an alternative site must be purchased
　Structure must be of a type of design/construction that allows it to be moved; for example, a wood-frame house is easier to move than a cinder-block house on a poured concrete slab

must be based on the natural character of the island, or what its character would be if not covered with buildings and asphalt. The next chapter outlines a holistic approach to mitigation.

When a storm strikes a developed barrier island, the typical reaction is to clean up the debris and rebuild. When Hurricane Frederic wiped out the Alabama-Mississippi coast in 1979 the community of Gulf Shores indicated that they were coming back "bigger and better." The motto for the recovery from 1992's Hurricane Andrew in Florida was "We Will Rebuild!"

Frederic cost $2.3 billion and 13 years later Andrew cost over ten times more at $25 billion. True, the comebacks are bigger as relief money spurs redevelopment: cottages are replaced by duplexes, duplexes by condominiums, low-rise by medium-rise if not by high-rise buildings. But better? Not really, because rarely do the "comebacks" result in planned communities based on island carrying capacity or planned islandwide mitigation of future property damage that *will* occur from future hazardous events.

The lessons nature is trying to teach us are going unheeded. The assumption is made that you can't do much about hurricanes anyway! But if something isn't done to mitigate property damage, one or two more tenfold increases in the cost of storm damage will break public treasuries, sink individual bank accounts, and erode the assets of insurance companies. Every barrier island property owner should be looking for new and better approaches to property damage mitigation. Every barrier island community should be working to reduce the risk faced by their population.

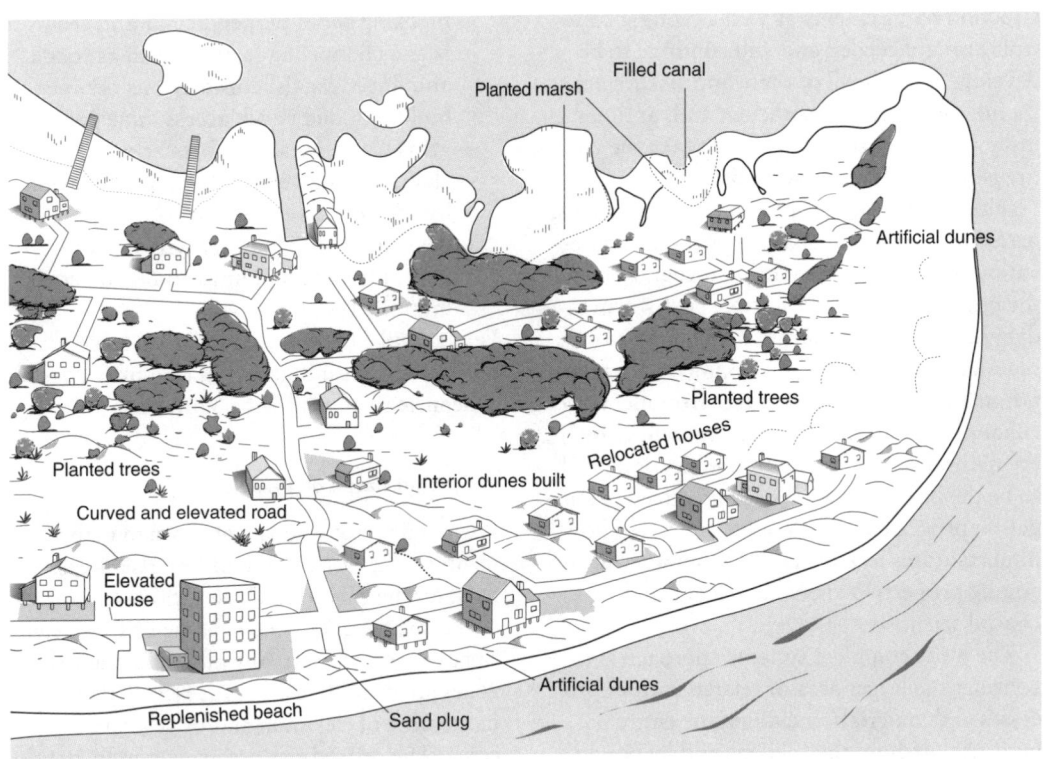

6.1 Pandora's Island after a lot of effort has been put into property damage mitigation. Artificial dunes have been put along the beach and interior dunes have been rebuilt. Roads run over and around, not through, dunes. Roads have been curved and new forest has been planted. The beach has been replenished.

How to Come Back "Better"

The "traditional" property damage mitigation options (see chapter 5), such as rebuilding frontal dunes using sand fences, stabilizing sand with vegetation, replenishing the beach, relocating damaged or threatened houses, and even armoring the shoreline under exceptional circumstances, should continue to be applied (see fig. 5.1). More rigorous design and construction requirements as well as zoning controls also are needed and will continue to be developed. While all of these approaches are useful, they remain insufficient and, at times, insignificant because they do not take the entire island system into account.

Initial preservation of natural environments, better recognition of coastal processes, conservation of sand and vegetation, recognition of the impact of historical storms, post-storm redefinition of coastal hazard areas, post-storm redesign of compatible development, augmentation and "repair" of island environments to enhance or restore protective capabilities of the natural setting, and public education need to be the basis for aggressive and effective mitigation programs. It cannot be overstated how important it is to increase the communication among property owners, managers, and coastal scientists.

The more complete systems approach is to consider the larger area of related natural processes and materials, including the entire island, the offshore shelf, inlets, and lagoon in a barrier island setting, or several city blocks inland for mainland ocean shoreline settings (e.g., the Grand Strand, South Carolina). The following discussion outlines property damage mitigation options from this more holistic perspective, including

—designation of inlet hazard zones (both potential and historical)
—building, augmenting, and repairing both frontal dunes and interior dunes

—blocking shore-perpendicular and cross-island channeling features such as roads and finger canals, constrictions between buildings, and beach access dune gaps
—grass plantings and reforestation
—change of road elevation and orientation by adding curves and rises
—development of long-term relocation plans for large buildings and entire communities (fig. 6.1).

For example, table 6.1 is similar to table 5.1 but emphasizes mitigation of the entire coastal zone instead of just the beachfront.

Mitigation Based on Mapping

Preliminary risk assessment through mapping identifies the distribution of low-risk areas, defined on the basis of high elevations and forest cover (see fig. 4.1). These characteristics need to be conserved. Virtually all remaining risk zones are candidate areas for remedial enhancement of elevation and/or vegetative cover. The revised risk assessment map identifies specific areas of the island that might benefit from mitigation efforts aimed at particular processes (e.g., storm-surge flooding, existing or potential overwash passes, potential inlet zones for new inlet formation or migration of existing inlets, back-island erosion zones, interior flood zones, deteriorating dune fields).

Evaluation of specific areas is usually tied to landforms and vegetation. Landforms (e.g., dunes) reflect the interaction of island pro-

cesses with materials (e.g., wind and sand), and, in part, vegetation (e.g., dune grasses). Remember that landforms are dynamic (e.g., dunes migrate) and are important in the island's response to storms (e.g., protective role of dunes against wave erosion, flooding, and wind in the shadow of the dune) as well as the island's migration (e.g., dunes may move along or across the island platform).

Vegetation almost always has a stabilization effect on the soil and topography from the trapping of sediment by grasses and mangroves to the anchoring of that sediment by dense mats of roots. Plant communities go through a succession when environments remain stable, from grasses to shrubs to trees that form the canopy of a climax forest that protects the diverse plants of the forest floor. Given the brief existence of barrier islands and the fact that they are primarily sand, the soils that support the plant cover are very fragile in the sense of nutrient organic content and texture. Removal of vegetation usually is accompanied by the loss of this thin soil and its remobilization by wind and water. Reestablishment of any vegetative cover is difficult, and even the hardy native plants adjusted to salt spray and the harsh environment of barrier islands will not reestablish quickly when the soil is disturbed. The construction phase for buildings and services often marks the beginning of ongoing problems for those same structures.

Working mitigation plans can be expressed

Table 6.1 Mitigation Options on Land

Abandonment

Relocation
  Active (relocate before damaged)
  Passive (rebuild destroyed structures elsewhere)
  Long-term relocation plans for communities

Soft Stabilization
  Adding sand to interior of island
    Rebuild interior dunes (including replacing roads)
    Raise island elevation (build artificial dunes and raised terraces)
    Infill ends of finger canals, or entire canal (potential new inlet)
    Infill road cuts
    Block interior cross-island roads (i.e., sand plugs, barriers, dead-end streets into forest)
  Vegetation (native species planting)
    Replace forest
    Plant tree/shrub thicket windbreaks
    Stabilize interior dunes
    Stabilize interior overwash terraces (grasses and shrubs)
    Plant marsh (*Spartina* and other natural marsh plants)

Modification of Development and Infrastructure
  Retrofit houses
  Elevate houses
  Reopen ground-level floors of elevated houses (that were enclosed)
  Curve and elevate roads
  Partially block or replace roads with interior dunes

Zoning, Land-Use Planning
  Recognize hazard areas and avoid. No construction:
    Tidal inlets (past, present and future)
    Swashes and breaches
    Critical environments (e.g., V zones, interior marshes, spits, shifting sand dunes, freshwater ponds)
  Choose elevated building sites
  Lower density development
  By ordinance:
    Protect interior dunes and other topographic highs against modification or removal
    Protect vegetation cover against removal or heavy disturbance

Things to keep in mind:
  Each island or coastal community is different.
  Consider entire coastal zone, not just oceanfront.
  Rising sea level must be considered.

---

as a mitigation map for an island or community in the coastal zone (see chapter 9). By way of specific examples, the major environments of typical barrier island systems are reviewed below and possible mitigation actions listed. Virtually every developed barrier island in America needs several or most of these types of mitigation applied if future property damage is to be controlled or reduced.

Island Interior Dunes

The importance of frontal dunes was discussed in chapter 5. Although frontal dunes were removed in the past, through mining or for a better ocean view, few would dispute the valuable role played in property damage mitigation by healthy, vegetated frontal dunes. Dunes found on island interiors also serve an important

Table 6.2 Field Trip Stops on Bogue Banks, NC, Illustrating Principles of Property Damage Mitigation

---

1. Triple ESS Pier. Beach Replenishment Project.
2. Old overwash pass—"effect of notching dunes." Also part of beach replenishment project.
3. Old development, filled-in salt marsh.
4. Preservation of dunes and new development, plus vulnerable construction in dune gap.
5. John Yancey Motel. Old and well-set-back development.
6. Dunes notched at ocean terminus.
7. Pine Knoll Shores Country Club. Artificial salt marsh shoreline.
8. Finger canal.
9. Beachfront condominium development.
10. Hurricane Hazel—former inlet site.
11. Hurricane Hazel—former inlet site.
12. Massive dune removal—early development.
13. Maritime forest and dunes removed, lawn planted.
14. Dynamic inlet site—Bogue Inlet.

Stop numbers correspond to circled numbers on figure 6.2.

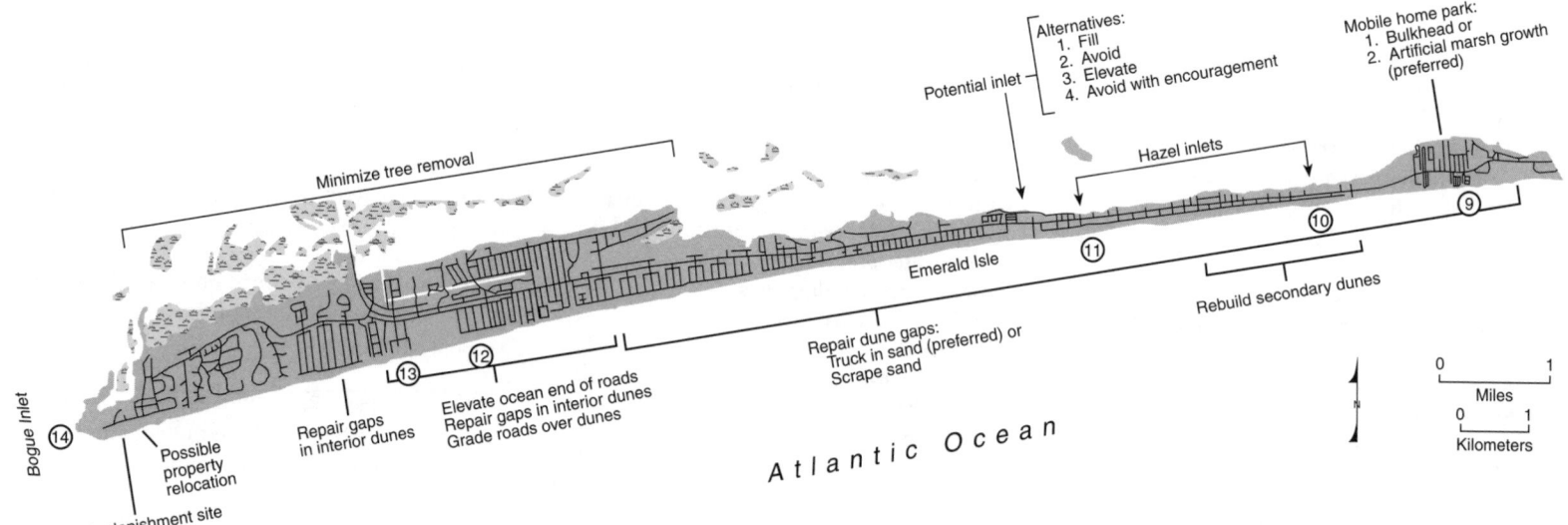

Minimize tree removal

Alternatives:
1. Fill
2. Avoid
3. Elevate
4. Avoid with encouragement

Potential inlet

Mobile home park:
1. Bulkhead or
2. Artificial marsh growth (preferred)

Hazel inlets

Emerald Isle

Rebuild secondary dunes

Bogue Inlet

Possible property relocation

Repair gaps in interior dunes

Elevate ocean end of roads
Repair gaps in interior dunes
Grade roads over dunes

Repair dune gaps:
Truck in sand (preferred) or
Scrape sand

Replenishment site

Atlantic Ocean

0 ___ 1
Miles
0 ___ 1
Kilometers

Bogue Banks (West)

6.2 A map illustrating recommendations for property damage mitigation on Bogue Banks, NC. This island has large areas of high dunes and heavy forest and is a good example of an island that could greatly benefit from relatively inexpensive property damage mitigation procedures. Numbers in circles represent stops on a field trip briefly outlined in table 6.2.

function with respect to property damage mitigation. Dunes provide elevation and protection (buffer) from storm-surge flooding and waves. Their height can act as a shield against damaging winds.

Frontal dunes are protected in many states, but interior dunes often are not given the same consideration. As a result, precious sand volume is reduced and elevation lowered, as has happened over and over again.

Bogue Banks, North Carolina (fig. 6.2), illustrates this problem. Near the western end of Bogue Banks are the highest sand dunes on the island and the largest in North or South Carolina south of Jockey's Ridge, North Carolina (number 13 on fig. 6.2; see also table 6.2). Large volumes of sand were removed from this location in order to provide a flat siting for residential development (see fig. 3.7). The

original dunes were very high and two rows were completely removed. The highest dunes, just to the west of the end of Ocean View Drive, are over 35 feet (11 meters) high. Here one can see the sharp bulldozed edge of the natural dune field and can get an idea of the magnitude of sand removal. In terms of storm damage mitigation, dune removal creates very serious hazards from floods, storm surge, and waves. Fortunately, onshore sand supply was great enough, and property construction here was set back far enough, so that sand fencing was effective in trapping sand and rebuilding dunes to afford some protection for property

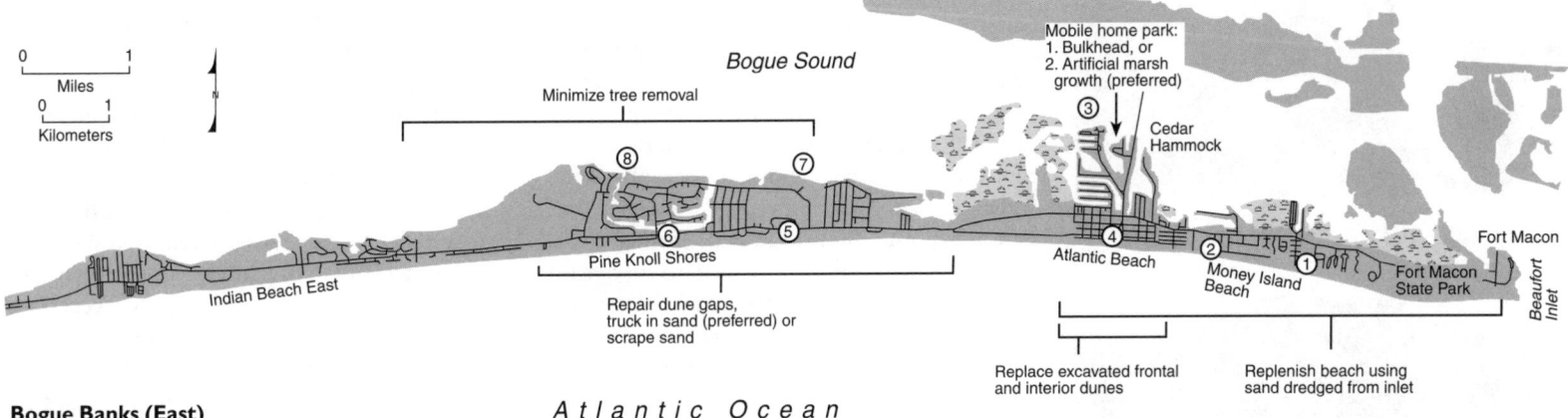

Bogue Banks (East)

Atlantic Ocean

owners from the threat of hazardous overwash. However, the new dunes are nowhere near the volume of the original dunes.

The lesson from Hugo in neighboring South Carolina was that the higher, wider, and more continuous the dune system, the greater the protection against damage. On Bogue Banks, continued sand fencing and even addition of sand from an off-island source are good options for frontal property. If a wide, healthy beach exists, front-island dunes have a sand supply for healing and growth. Interior dunes on developed islands do *not*: the sand supply is covered by houses, lawns, and streets. Once

an interior dune is destroyed, it can only be rebuilt or repaired artificially. The better choice is to conserve the dunes and design the architecture around and over the landforms. Damage potential will be less, the costs of dune reconstruction eliminated, and a safer, more aesthetically pleasing development will result. Table 6.2 provides a brief description of several locations on Bogue Banks, shown by circled numbers in figure 6.2, that illustrate some property damage mitigation principles.

The central portion of Galveston Island, Texas, was elevated in response to the tragedy of 6,000 deaths in the 1900 hurricane (fig. 6.3;

see also chapter 9). Though not formed into distinct dune ridges, this is another method of adding elevation to islands and thus reducing floodability and potential for property damage.

Coastal Vegetation

Natural coastal vegetation, where little disturbed by development, offers some of the best defense against property damage during storms. For very large storms, no amount of vegetation or dense forest cover may be sufficient. For many moderate-sized storms, how-

6.3 After the devastating 1900 hurricane, the elevation of the central city of Galveston, TX, was raised by pumping sand into the community. Some buildings were raised, but for others first floors became basements and second floors became first floors. This was a wise move to reduce damage in the next storm. Photo furnished by Robert Morton.

6.4 A house at high elevation surrounded by forest on Pawleys Island, SC, immediately after Hurricane Hugo. This house survived relatively unscathed because of its excellent siting.

ever, dense forest, especially native species of maritime forest, provides significant protection to buildings.

Maritime Forest

Maritime forests usually grow only on stabilized dune systems and generally on the back sides of islands. Although true maritime forests grow in areas exposed to the ocean, they flourish where island width, topography, and orientation provide sufficient protection from storm exposure (Bourdeau and Oosting, 1959). Such forests have evolved to survive under the harsh conditions found within the coastal zone, including salt spray, wind shear, nutrient-poor soils, and low water availability (Barbour, De Jong, and Pavlik, 1985).

The protective nature of maritime forest was well illustrated when Hurricane Hugo hit Pawleys Island, South Carolina, in 1989 (fig. 6.4). Overwash penetration and storm wave damage to property was noticeably greater where maritime forest had been removed for development. Neighboring houses suffered vastly different degrees of damage from Hurricane Hugo. Many houses located within the maritime forest were essentially untouched except for some cosmetic damage. Many houses built in cleared areas were destroyed. Dauphin Island, Alabama, provides a similar example. Hurricane Frederic in 1979 did extensive damage to the unforested western segment of the island, while the forested eastern segment suffered less damage. That pattern of destruction on the island was repeated in later events.

The densely forested, higher elevation areas are the most stable and lowest risk portions of barrier islands during a storm. Every measure should be taken to protect and preserve forest growth. Other aspects of island vegetation are also noteworthy and discussed below. Devegetation of the coastal zone only increases the susceptibility and likelihood of storm damage. Not only should as much forest as possible be retained, but also, where appropriate, areas where trees have been removed should be reforested with native species. Once newly built dunes are stabilized with grassy vegetation, forest growth can be encouraged.

The town of Pine Knoll Shores, Bogue Banks, North Carolina, is one of the most enlightened communities with respect to sound development practices for living with the shore. Town restrictions dictate that a permit is needed to cut down any tree over 2 inches in diameter. The community also controls density of development, attempting to preserve the groundwater as well as minimizing destruction of vegetation. It helps that much of Pine Knoll Shores is on a naturally elevated part of the island. Their secret is developing on the lower risk portion of the island and protecting the natural amenities that make it less vulnerable as well as a more pleasant place to live.

Bogue Banks also provides common examples of the "bad," including extensive clearing of maritime forest (as well as leveling of interior dunes). In one example, near the western end of the island (see fig. 6.2, stop number 12), about 40 acres of maritime forest were removed and sand dunes were leveled for the siting of a large motel complex (fig. 6.5). This occurred before laws were in place protecting frontal dunes. These buildings sit on top of what amounts to a sandy bluff, 12 feet (3.5 meters) or so high, eroding at a long-term average rate of 2 feet (0.6 meters) per year. In addition, several nearby roads run perpendicular

6.5 A diagram showing the site of the Islander Motor Inn in Emerald Isle on Bogue Banks, NC. Here more than 40 acres of very dense maritime forest were removed to put in the motel. Dunes were flattened and replaced by lawns. An area with low property damage risk was altered to one of high risk.

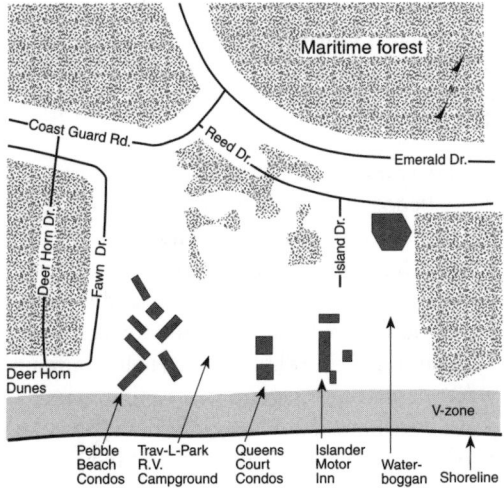

to the shore, and the dunes are notched allowing for increased overwash penetration and storm-surge ebb flow.

There really was no need to clear so much forest for the development. The motel is left with a large, flat lawn instead of natural nearshore vegetation. The lawn will cost more to maintain and will offer no protection from storms. Recommendations for this area are to rebuild the frontal dunes and the interior dunes as extensively as possible. The maritime forest should be reestablished, although that is a much longer term project. Ocean ends of roads should be closed (plug dune gaps) or curved and elevated where possible.

Mangroves

The dense mangrove forests of the south Florida mainland helped reduce the coastal impact of Hurricane Andrew in 1992. Mangroves serve as a primary sediment trapper and anchor for tropical/subtropical barrier islands and low-lying shores. When mangroves are removed for development, to provide boat access, or simply to provide an unobstructed view, shoreline erosion greatly accelerates (fig. 6.6).

The protective effects of mangroves and coastal vegetation in general were noted on the coast of the Yucatán Peninsula, Mexico, after Hurricane Gilbert in 1988. South of Cancún, the eastern (Caribbean) coast of the Yucatán is largely undeveloped except for the

6.6 (a) A typical undisturbed mangrove forest dampens wave energy and causes sediment accumulation just like a salt marsh. (b) Once the mangrove has been removed, protection is lost and shoreline erosion begins. These examples are from La Parguera, on the southern coast of Puerto Rico.

towns of Puerto Morelos and Playa del Carmen. Here is a striking example of the protective value of natural, undisturbed grasses and forest. In zones of undisturbed vegetation (see fig. 3.18a) the penetration of overwash sand was limited to a few tens of meters except along the one shore-perpendicular road. In the development in Puerto Morelos, however (see fig. 3.18b), where the natural vegetation and dunes were removed, overwash sand penetrated the width of the development, a distance of at least 200 meters.

Lagoonside Marsh Growth

Salt marsh grasses, typically *Spartina*, are efficient trappers of fine-grained sediments in the shallow lagoons and tidal mudflats behind barrier islands and along embayments. The extensive salt marshes behind America's barrier islands trap tons of sediment before it reaches the ocean, providing the substrate and nutrients that sustain the associated ocean fisheries for both shellfish and fin fish. Marsh grasses thrive on burial and trap muddy sediment by baffling wave and current energy. A marsh meadow essentially eliminates wave energy. When marshes die or their grasses are removed, rapid erosion begins immediately. Reestablishing marshes is a very effective way to combat erosion on the back sides of barrier islands and other low-energy saltwater shores.

Planting salt marsh may be an environmentally sensitive alternative to bulkheading in order to combat lagoonside erosion of islands (Broome, Woodhouse, and Seneca, 1982a). A salt marsh was cultivated to control erosion on the lagoon side of a central portion of Bogue Banks, North Carolina (see fig. 6.2, stop number 7), at the Pine Knoll Shores Country Club (fig. 6.7). A salt marsh was successfully culti-

vated here to stabilize a shoreline that was eroding at a rate of more than 20 feet (6 meters) per year (Stanczuk, 1975). A narrow strip of salt marsh, only a few feet wide, was planted around 1973. Today the marsh is over 100 feet (30 meters) wide. The salt marsh acts as a buffer to wave action and is a simple way to build up the lagoonside shore and reduce the effects of a major storm. Artificial plantings of salt marsh grass to establish protective marshes is a common mitigation practice for low-energy shorelines. Chesapeake Bay shows that success of marsh planting as a stabilizing barrier is partly dependant on fetch. Planting is most successful where less than one mile of open water is available for wave generation. Large fetch allows large waves during storms to prevent the establishment of marsh.

Back-barrier marshes help to contain flood waters, dampen lagoonside wave energy, and add width and elevation to the island by trapping sediment. The back-island shores, however, may be bounded by tidal creeks and channels or artificial channels such as parts of the Intracoastal Waterway. Where marsh is absent, the back sides of islands tend to be erosional. Bulkheading has been the response to lagoonside erosion. Yet this ultimately decreases the island's width, and in a scenario of rising sea level could weaken an island's defenses (e.g., as at Topsail Island, North Carolina).

One of the main processes by which an island migrates in response to sea-level rise is by

6.7 This marsh on the sound side of Bogue Banks, NC, is adjacent to the golf course at the Pine Knoll Shores Country Club. When the golf course was constructed in the early 1970s, trees along the shoreline were removed and the erosion rate jumped from 1 foot per year to 25 feet (0.3 to 7.5 meters) per year. The golf course owners planted a narrow strip of salt marsh grass, which in 20 years grew to a 200-foot-wide (60 meters) healthy marsh with essentially no shoreline retreat. Unfortunately, this example of shoreline stabilization is rare. Most property owners prefer the quicker fix of a bulkhead.

sand overwashing the island during storms and being deposited in the lagoon. Bulkheading of the lagoon side of an island tends to prevent overwash sand from reaching the lagoon by interrupting the sediment transport and keeping the lagoon shore from migrating toward the mainland. Bulkheading tends to help maintain water depth on its lagoon side so that marsh grass does not reestablish growth. Bulkheading may also cause wave reflection and increased scouring, enough to erode or discourage marsh grass growth. As erosion continues on the front side, the island will tend to narrow.

The island narrowing problem is real, though there is no way to predict the time frame over which it will occur. It could be centuries, or perhaps only decades, depending on the sea-level rise, storm climate, sediment supply, and other variables. Even some mitigation recommendations might exacerbate the problem of island narrowing. For example, methods to reduce overwash certainly prevent the

island from natural migration (and elevation increase) in the long term. A method that is used in some places to combat narrowing is to replenish the lagoonside shore of the island, in effect acting as artificial overwash. Planting marsh grass and conserving existing marsh are alternatives to bulkheading.

Orientation and Placement of Roads and Services

Given the damage patterns from Hurricanes Gilbert (1988) and Hugo (1989) and other storms, it is clear that road orientation, design, and placement usually are major factors contributing to increased storm damage. Some of the problems stemming from standard road development plans and practices can be remedied or at least ameliorated. Most mitigation mea-

sures concerning development are along the lines of engineering and building codes. Buildings and services may be built stronger, but their placement needs to be consistent with the natural processes to which those structures are subjected. Such considerations should be addressed by the engineering community. For example, roads, streets, water lines, and other utilities are laid out in the standard inland grid pattern. Buildings block natural flow (e.g., overwash) while the ends of streets and gaps between rigid buildings funnel and concentrate flow, accentuating the erosive power of flood waters.

During Hurricane Hugo, for example, water, sand, and debris were carried inland along shore-perpendicular roads in several South Carolina communities. On the northern end of Pawleys Island such roads acted as storm-water conduits, which led to a great deal of property damage. In addition, Pawleys' boat ramps provided ideal conduits for the return of storm waters (storm-surge ebb) back onto the island from the lagoon. Storm-surge ebb caused scour channels, which undermined roadways and damaged adjacent houses and property. Such damage patterns were widespread after Hugo and in Mexico after Hurricane Gilbert.

If the direct line created by straight roads perpendicular to the shore could be interrupted, the amount of damage done by overwash and storm-surge flood and ebb waters could be reduced. Exactly how to block

6.8 The orientation of these roads in Volusia County, FL, apparently was due to the orientation of the original township and range boundaries. Whatever the reason for their orientation, angling the roads away from the beach may reduce damage in future storms.

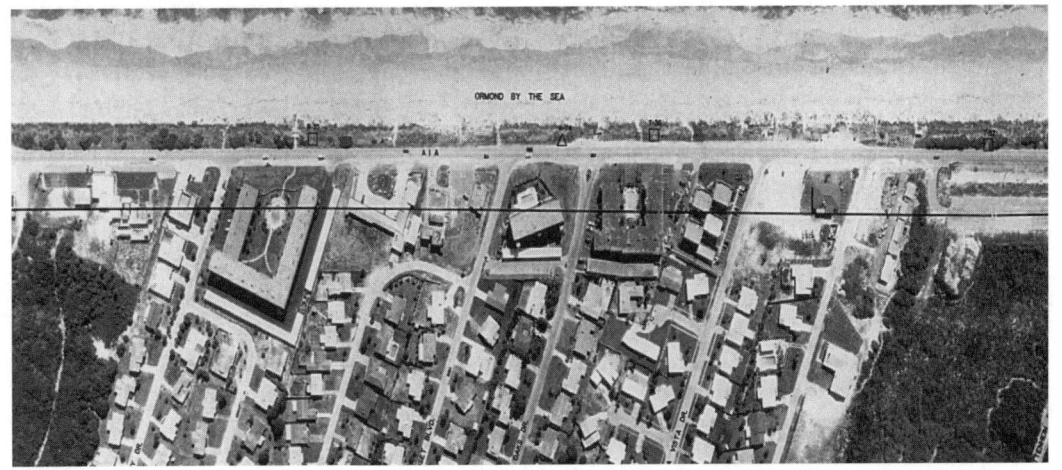

roads in order to reduce the potential for overwash and/or storm-surge ebb may generate controversy. Obviously, if major gaps exist in the frontal dune ridges, the gaps could be plugged as described below. Simply building a small "bump" on the oceanward terminus of the shore-perpendicular roads could at least slow down the storm-surge ebb velocity enough to reduce the scouring potential of the flow and reduce or delay the intrusion of storm waves into the community. Such a design is presented in the Texas General Land Office booklet (1991; see page 18, figure 30 of that report).

Just adding a few simple curves in roads, instead of having all access roads running perpendicular to shore, would greatly reduce the impact of overwash and storm-surge ebb. In

6.9 At the Land's End subdivision, Emerald Isle, Bogue Banks, NC, the roads are curved, go over rather than through the dunes, and do not terminate on the beach. This stands in great contrast to the development shown to the right of the subdivision, with shore-perpendicular roads extending to the beach. Land's End will suffer much less property damage in a future major storm.
6.10 Shown here is an artificial mound of sand, suggested as an approach to prevent overwash and wave attack on a low-lying beach community. This approach to property damage mitigation is much longer lived and much cheaper than a replenished beach. Obviously such an approach requires a changing of traffic patterns within a community. Drawing by Terese Capal.

Volusia County, Florida, east-west roads parallel township and range lines and are thus at some angle to the shoreline (fig. 6.8). This was probably a simple coincidence of "laying out the city," but will probably help lower property damage during a major event. The Land's End subdivision near the western end of Bogue Banks, North Carolina, has a very curved road layout (fig. 6.9); while it is likely this was done for aesthetic reasons, it certainly will help mitigate property damage in the next storm.

Typically, more recent developments and planned island communities such as Kiawah and Hilton Head islands, South Carolina, avoided shore-perpendicular roads. Older communities must consider relocating roads, adding curves, or otherwise interrupting the grid pattern so as to eliminate the conduit effect. Placement of T-mounds—simply, T-shaped artificial sand dunes (fig. 6.10)—at alternate intersections is effective and does not place any demand on private property (as road relocation might). Traffic flow is reduced in neighborhoods, and the T-mounds can be designed to function as miniparks or playgrounds or for aesthetic plantings. T-mounds can serve a local function of protective dune. Similar sediment mounds can be constructed around other likely flood conduits, for example, at the ends of finger canals or on the back side of the island in the vicinity of boat ramps.

Road orientation and placement can have a major impact on interior areas flooded,

overwashed, and subject to storm-surge ebb. They will not, however, reduce frontal wave impacts or have a significant mitigative effect against waves in the V zone. Although services and utilities do not contribute to erosive processes like roads do, their presence can create the mentality of a "line in the sand" that must be protected or held. Instead, their placement can be planned with hazards in mind. For example, buried lines may be less subject to wind damage; however, burial must be deeper than the anticipated depth of scour and out of any erosion zones. Trunk lines should be in the interior of the island.

Road Elevation

Every community on Topsail Island, North Carolina, should prohibit the construction of any new structures seaward of the main road. A particular problem is the population density on the northern end of Topsail, which is at a critical point. State Route 1568 is often overwashed along a wide front, cutting off the only evacuation route from the northern end of the island. Some spots on this solitary road are overwashed even in moderate storms and are protected only by a small, essentially inconsequential sand ridge that is artificially maintained by bulldozing (fig. 6.11). Moreover, the newly built section of the road, where it has been relocated landward to the back side of the island, is in danger from flooding because of its low elevation.

When roads are located, or rebuilt/relocated after storms, they should be elevated

6.11 The escape route for one of North Carolina's densest shoreline developments, the north end of Topsail Island. Property owners face not only serious propery damage risk, but also the threat of not being able to escape a storm unless they leave early.
6.12 This view of Highway 12 on Hatteras Island, NC, shows mounds of overwash sand on either side of the road that was removed after storms. Overwash sand is an important aspect of island migration, as it furnishes elevation to the island as it migrates in a landward direction. Frequent flooding of Highway 12 is a major evacuation hazard.

for several reasons. Most fundamental is that the road should be above the initial flood level of a storm in order to allow for evacuation. The added elevation may offer some protection against storm-surge currents, if not flooding. However, road elevation design must pay careful attention to natural processes so as not to interfere with natural drainage or sediment transport. This applies as well to overwash sediment, described in chapter 3 as nature's way of building up island elevation as well as moving an island landward during a rising sea level. If the sand is removed every time a road is buried by overwash, then island elevation cannot build up. This management strategy is used along N.C. Highway 12 on Hatteras Island and has resulted in Highway 12's being in a "furrow" along part of its length (fig. 6.12). Wouldn't it be better to let island elevation build up naturally, sacrifice the paved section of road buried by overwash, and replace those short sections of paved road with (very inexpensive) gravel road? Sure, it would mean having to drive at 35 miles per hour instead of 55 (or faster), but what an incredibly inexpensive, environmentally sensitive option!

These same concepts apply when designing bridge approaches and access causeways to islands. The causeway to Figure Eight Island, North Carolina, blocked the tidal flow in the salt marsh crossed by the causeway, resulting in loss of marsh and impact on the protective environment of the back side of the island. Maintaining the natural flow and currents in the marsh wetlands should be part of any construction design.

The Complexity of Barrier Islands: Island Morphology

Modern barrier islands rim most of the U.S. Atlantic and Gulf of Mexico coasts. These relatively young geologic features are only the most recent of several generations of barrier islands that have existed in the same or similar locations over recent geologic time. Sea level has risen over the continental shelf countless times, each time bringing with it another generation of barrier islands. Each series of islands were left stranded as the sea, inevitably, began to recede. Depending on the maximum level of the sea, some islands were at higher elevations and more inland positions than the present ones, and some were at lower elevation and more seaward positions. The coastal plains are filled with remnants of ancient barrier islands. More significantly, many present-day barrier island complexes include portions of these older barrier islands, and some mainland shorelines are formed along these older, stranded barrier islands (e.g., the Grand Strand, South Carolina).

Ancient barrier islands can play an important role in the relative risk of a modern island with respect to potential property damage, as is the case in Kitty Hawk, North Carolina. Here, a remnant of the ancient barrier island remains to form the back side of the island complex. The higher elevation, forested portion of Kitty Hawk was left stranded by an ancient, higher sea level (see fig. 4.14). The modern island (today's beach, dune, and overwash fan system) has been welded onto the ancient island, and the island complex illustrates how the island's geologic history controls landforms, processes, and materials and forms a basis for a rational management program.

Kitty Hawk is an old Outer Banks community, first settled in the mid-1800s. The town is bordered by Southern Shores to the north and Kill Devil Hills to the south. Predominantly a tourist community, Kitty Hawk has approximately 1,700 full-time residents, with the population swelling to well over 15,000 during the summer months. The island is over 2.5 miles (4 kilometers) wide with significant areas of high-elevation dunes, mostly on the back half of the island along Albemarle Sound.

The primary dune along the beach of Kitty Hawk ranges in height from less than a foot to 20 feet (0.3 to 6 meters) with approximately two-thirds of the beachfront possessing dunes below 10 feet (3 meters) in height. The original dune was constructed in the 1930s as part of the huge artificial dune line extending from the Virginia border to Ocracoke, North Carolina. Today, where a dune does exist, often in front of lots with no structures, gaps in excess of 10 feet (3 meters) are common and the dune is usually unvegetated and contains shell material, indicating that the sand was bulldozed. The 3-mile (4.8-kilometer) stretch of Kitty Hawk ocean shoreline is experiencing average erosion rates of approximately 5 feet (1.5 meters) per year (NCDCM, 1992).

The interior portions of Kitty Hawk are characterized by low, flat, partially vegetated land, and this region is extremely vulnerable to overwash, storm-surge flooding, and wave action during storms. Significant vegetation exists in places, including live oak, pine, grass, and shrub thicket. The dominant feature of the island's interior is Kitty Hawk Woods, a combination of dense forest and marsh vegetation on Roanoke Sound. Elevation ranges up to well over 20 feet (6 meters).

Development along the beach of Kitty Hawk consists predominantly of small, 30+-year-old wooden homes, most elevated between 6 to 10 feet (1.8 to 3 meters) on wooden pilings. A large number of these homes are located shoreward of the primary dune field, within 100 feet (30 meters) of the high-tide line. Figure 4.14 shows typical Kitty Hawk settings and development. Dozens of buildings have either been lost to erosion or moved back during the last two decades.

Figure 2.2 shows that in a typical modern barrier island, the front (ocean) side has the highest elevation because frequent overwash events add sand and elevation to the island. The front side is also where the strongest winds blow, building up loose sand to form dunes. Usually, island elevation decreases from the ocean toward the lagoon, though dune field width and distribution may interrupt a simple profile.

In contrast, the result of the two islands merging together is a high elevation back-side of the island (the ancient island core) and a high elevation front-side of the island (the modern overwash apron and dune field), with the lowest island elevations found in the island interior between the dune fields. In the case of Kitty Hawk, two other factors are important: First, the ancient island core is much higher in elevation than the modern, so the highest island elevations occur on the very back side of Kitty Hawk; second, the frontal dune in Kitty Hawk (and along the entire Outer Banks of North Carolina) was artificially constructed during the 1930s. This dune has blocked the overwash process from occurring for some 60 years, which seems like a good thing for protecting the island from flooding. However, the real end product is an island that has been *cut off* from its elevation-increasing process for decades, resulting in an island that is now lower than it would be otherwise. Today, the artificial dune line is breached in places and about to be breached in others, meaning smaller and smaller storms will be able to flood the island. Where will the flood waters go? Right into the "bowl" that is the lower elevation central portion of the island.

The likely future for Kitty Hawk (and other Outer Banks communities) is that the flood frequency will increase as the dune continues to be destroyed and the dune breach widens. Here is a case where some oceanfront homes may have to be sacrificed in order to have room to rebuild a dune to protect the island's much more extensive interior property. Of

6.13 A typical finger canal in Myrtle Beach, SC, provides waterfront property for a large number of property owners. The problem with finger canals is that they also provide access for storm-surge floodwaters, provide potential sites for new inlet formation, and create groundwater contamination problems on the island. Photo by Gered Lennon.

course, this solution perpetuates the problem of cutting off the overwash source of island sand. Artificially building up the island's interior elevation will be difficult, given the density of development, but such an option shouldn't be ruled out (see the Galveston Island, Texas, response to the 1900 hurricane, fig. 6.3).

Finger Canals

A common man-made island alteration that causes a variety of problems is the finger canal (fig. 6.13). *Finger canal* is the term applied to the ditches or channels dug from the lagoon or sound side of an island into the island proper for the purpose of maximizing the number of waterfront lots. Canals can be made by excavation along or by a combination of excavation and infill of adjacent low-lying areas (usually marshes). The resulting substrate is not the best in which to anchor building supports.

The major problems associated with finger canals are

(1) the lowering of the groundwater table
(2) pollution of groundwater by seepage of salt or brackish canal water into the groundwater table
(3) pollution of canal water by septic seepage into the canal
(4) pollution of canal water by stagnation due to lack of tidal flushing or poor circulation with sound waters
(5) fish kills generated by higher canal-water temperatures
(6) fish kills generated by nutrient overloading and deoxygenation of water

Bad odors, flotsam of dead fish and algal scum, and contamination of adjacent shellfishing grounds are symptomatic of polluted canal water. Thus, finger canals often become health hazards or simply places near which it is too unpleasant to live. Residents along some older Florida finger canals have built walls to separate their cottages from the canal! If the canal acts as a sediment trap, its function as an access navigational channel may be reduced over time.

Finger canals have another destabilizing effect. Where canals cut deep into the island's interior, almost to the ocean side, storm-surge ebb flow may be funneled through the canal and cut an inlet, or adjacent finger canals may lead to lateral breaching, creating small backside islands (e.g., Alabama after Hurricane Frederic in 1979; see fig. 1.3). Numerous finger-canaled barrier islands will be breached along these zones of weakness in future storms, creating greater risk during the storms and expensive post-storm island restoration. Pre-storm mitigation is recommended.

As previously noted, building mounds at the landward end of finger canals or other potential points of breaching is a simple method of mitigation. Shortening existing canals by infilling their heads is a more costly approach but may be warranted. In new developments, finger canals should be permanently banned. Sometimes there are trades, that is, old canals filled if new ones are created. Check your state and local management office. When inlets do form due to finger canals, the post-storm reconstruction offers the opportunity to fill in both the inlet and the offending canal so as not to repeat the problem. Chapter 7 discusses the hazard set that is specific to inlets.

Swashbuckling

Somewhat analogous to inlets cut through finger canals are the surge channels eroded across the island's interior topographic lows, either natural or man-made when development plats are excavated and leveled. These shallow waterways form during storms or from freshwater runoff and are known as *breaches* or *swashes*.

The Grand Strand, South Carolina, area is a mainland coast that has a series of swashes instead of inlets at ends of barrier islands. Swashes are small, in some cases intermittent, streams draining fresh water from the mainland. They are also very low elevation areas and particularly prone to flooding, storm overwash, and storm-surge ebb funneling. Development near swashes places property and infrastructure in danger. In addition, the main shore-parallel roads of the Grand Strand area pass over these swashes on small bridges.

Swash locations are easy enough to spot. Even though they may be dry or have only intermittent freshwater flow, during storms they act like sluiceways. To reduce property damage, no more development should take place in these hazardous zones. Swashes and storm breaches should be treated more or less like inlets on barrier islands (see chapter 7). The swashes will migrate or meander just as typical rivers and streams do. Property located behind the swash, even if well inland, is also in danger because of the low elevation of the swash area and the potential of swashes being exploited by storm surges.

To completely fill the swashes is not reasonable because channels for storm-water drainage must be provided. Replacement of some of the swashes with storm sewers is a possibility, but such systems sometimes cause flood waters to back up, raising flood level. Before modifying these natural drainage outlets, input from an urban storm-water hydrologist and/or engineer should be sought to determine the impact on adjacent properties, including inland areas.

Hazards under Construction

Much of the damage to barrier island landforms and forest occurs during the construction phase of a building and can be prevented. Even for just a single-family house, land is leveled, dunes notched, and vegetation removed. These changes are usually meant to be temporary, but in reality, the damage is done, the risk greatly increased, and reconstruction is difficult. Buyers should work closely with architects and contractors to design and build in conformity with existing topography and natural vegetation. The class developments on barrier islands are the ones that have kept dunes and forest intact (e.g., Kiawah Island, South Carolina; interior developments on Hilton Head Island, South Carolina; Pine Knoll Shores, North Carolina). The aesthetics are better, the resale value is higher, and best of all, the property is more likely to survive until its resale date!

Offshore Rubble

Landward transport of frontal houses and associated debris during storms is another type of development hazard. Buildings are carried into island interiors or across the island into the marsh—unless another house is encountered and smashed along the way. Less well known is oceanward transport of debris off of the island. Gayes (1991) in a side-scan sonar survey taken immediately after Hurricane Hugo found a considerable amount of debris that was carried offshore by storm-surge ebb. In addition, he found that storm-surge ebb waters were funneled by development, carving channels into the nearshore environment. Small "deltas" of sediment and debris also were observed directly offshore of the swashes along the Myrtle Beach area. Rubble carried offshore into the ocean, inlet, and marsh during a hurricane is a real hazard to swimmers and boaters. A Potential Debris Inventory should be prepared immediately after major storms so that potential hazards can be identified and removed as part of the post-storm reconstruction.

Manufactured Housing

Manufactured housing presents a special case in terms of the construction-related hazard. Several coastal communities have very high densities of mobile homes. For this reason alone they would suffer extensive damage in a major storm. Mobile homes provide a different problem with respect to mitigating property damage, and their presence should raise a red flag for community planners and nearby prop-erty owners. Because of their vulnerability to wind damage, locating mobile homes in wooded areas and conserving protective vegetation is even more important. Their light weight makes mobile homes susceptible to overturning and floating, creating a hazard to adjacent units and other structures (e.g., missiling and ramrodding) (fig. 6.14). Mobile homes must be tied down as securely as possible. Again, trees and dense shrub cover between housing units may offer additional protection. In many instances mobile home parks are located on the lagoon side, on marsh fill (no longer legal), and at very low elevation, making them susceptible to flooding. Such parks should not be allowed to intermix with permanent dwellings, but facilities similar to those for day campers and RVs that can be evacuated might be permitted. The use of low-elevation coastal areas for RV camps north of Myrtle Beach, South Carolina, is a good example. Again, avoid grid street patterns in mobile home parks.

Case Study: A Plan for Folly Beach, South Carolina

Folly Island and the community of Folly Beach were introduced in chapter 3 in the context of the impact of Hurricane Hugo. The island provides a good example of some specific hazards and ways to mitigate such storm damage. Although these approaches are suggested specifi-

6.14 Mobile homes are highly susceptible to damage in storms. Shown here is an overwash fan made up of mobile homes on Myrtle Beach, SC, after Hurricane Hugo. Photo by Rob Thieler.

cally for the community of Folly Beach, they can be utilized anywhere.

The general theme is that the community needs to develop a long-term plan to roll with the shoreline migration, but in an organized fashion. Building elevation, particularly through building dunes and other sand additions, and reestablishing a protective vegetation cover are ongoing objectives, but at the same time, small-scale projects and local ac-

tions can reduce potential losses immediately. Relocating buildings must also be incorporated into the intermediate time frame of the plan. Clearly, roads running perpendicular to the shoreline increase the potential for overwash, local storm-surge flooding, and storm-surge ebb scour. The effects of these processes could be reduced easily and relatively inexpensively. The proposed method is to add T-mounds, artificial dunes, or other natural obstructions at critical points to block some of the paths for storm surge and overwash, as well as blocking paths that storm-surge waters would exploit in flowing back to the sea. Simply blocking the conduit effect of some of the roads running perpendicular to the shore would slow and dissipate the future return flow or channel the water through undeveloped zones. The likelihood of increased flooding caused by ponding of flood waters is minor compared to the potential surge damage, and flooding will occur anyway unless the entire island is raised in elevation. Surge waters would be slowed enough to flow back to sea at a rate that would not cause scour, but not slowed enough to increase flooding significantly. Such a solution needs the design input of a hydraulic engineer.

Selective addition of sand should include constructing a series of artificial sand dunes wherever possible on the island to add as much sand volume as practical and to block up as many of the shore-perpendicular (and in some cases, shore-parallel) roads as possible. Figure 6.15 shows such a hypothetical dune, 30 feet

(9.1 meters) wide at its base, 10 feet (3 meters) wide at the top, 10 feet high, and 200 feet (61 meters) long. The volume of sand contained in the dune is 1,500 cubic yards (1,100 cubic meters). A reasonable higher-than-average cost of emplaced sand in South Carolina is $5 per cubic yard. As such, this dune would cost about $7,500. Again, sand placed on the island is "permanent." The same volume of sand placed on the beach for beach replenishment would be almost insignificant, as well as being removed quickly, probably by fairweather waves, before it had a chance to afford any protection against storm waves. Such island interior dune construction projects are independent of shoreline mitigation, but should be given serious thought for inclusion in all future beach nourishment projects when large sand volumes are imported, and the equipment is in place for moving sand, thus reducing costs.

Changing the orientation of some streets so that they do not run directly perpendicular to

6.15 Diagram showing a hypothetical artificial dune constructed from 1,500 cubic yards of sand and costing between $7,000 and $8,000.

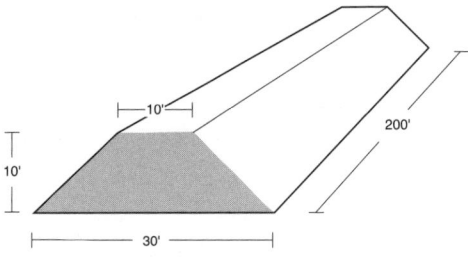

shore will likely decrease some of the effects of overwash and storm-surge ebb. Figure 6.16 shows three site-specific plans to reduce the impact of overwash and storm-surge ebb on Folly Beach. The circles represent structures (mostly single-family homes). The shaded areas are dunes, each of the same volume as shown in figure 6.15 but in the shape and orientation as shown in figure 6.10. The three examples in figure 6.16 show approximately the same amount of dune building, but increasing amounts of street re-orientation. One important restriction placed on the proposed design is that no buildings were to be moved.

Relocation (see chapter 5) is an option gaining popularity, but it is often difficult to get public support in situations where houses cannot simply be moved straight back on an owner's property. Folly Beach is so densely developed that in order for relocation to work, people would have to move their homes to other parts of the island or off the island. Recall from chapter 5 that demolition and rebuilding elsewhere is also considered a form of relocation. Economics, construction design, and lot size dictate the preferred method.

Figure 6.16a shows simple dune building and blocking of intersections with no buildings moved and access still easy for all residents. Figures 6.16b and c show increasingly invasive forms of street blocking and re-orientation. Figure 6.16b depicts minor road building and moving of parts of roads, but basically it suggests simply blocking and curving road inter-

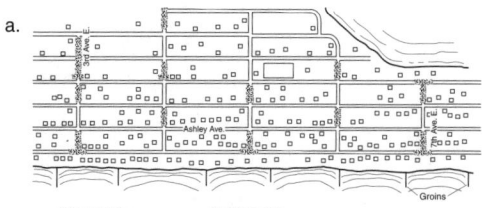

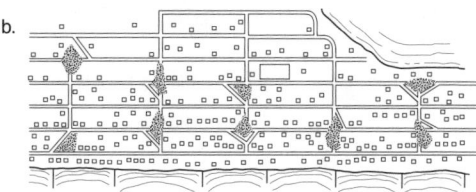

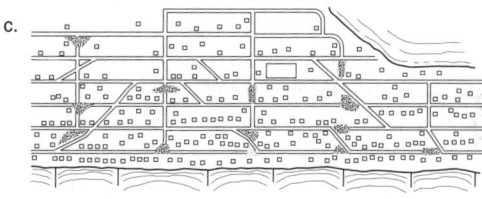

6.16 Three plans to reduce the impact of overwash, storm surge, and storm-surge ebb in a future storm on Folly Beach, SC. The three plans depicted use approximately the same amount of sand for dune building, but use increasing amounts of street reorientation from a to c. No buildings (small squares) are moved.

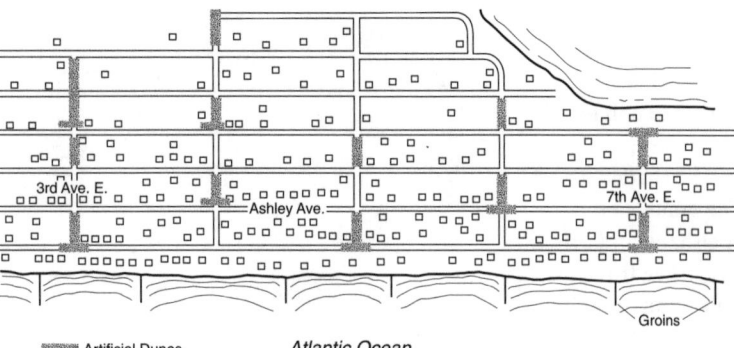

Artificial Dunes   *Atlantic Ocean*   Groins

3rd Ave. E.   Ashley Ave.   7th Ave. E.

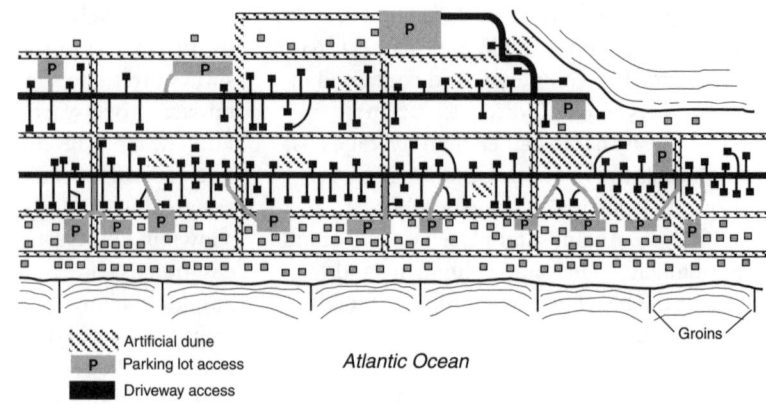

Artificial dune
P   Parking lot access   *Atlantic Ocean*   Groins
Driveway access

sections to cut down the number of straight through passes for storm waters. The artificial dunes are placed as shown by the shaded areas. Points of placement should consider where some reduced ease of access would be considered beneficial (e.g., less through traffic), but all residents would be able to drive to their homes simply by rerouting around an additional block or so. Figure 6.16c represents the most invasive (and thus most costly) plan, where many of the streets are re-oriented, necessitating building of streets on presently empty lots. The lots would need to be purchased by the town, donated, or traded for other land. Still no buildings would be moved. The T-shaped interior dunes (see fig. 6.10) used to block intersections should be stabilized either with vegetation (native species) or with an erosion-resistant surface if designed

for a secondary use (e.g., playground, park, overlook).

Figures 6.17 and 6.18 show additional plans for blocking overwash and storm-surge ebb paths by artificial dune building. In the plan shown in figure 6.17, there would be little or no decrease in access, as the "buttress-shaped" artificial dunes block only a few intersections. The T-shape is a plan to get as much sand as possible perpendicular to the shoreline. Figure 6.17 shows 13 of the artificial dunes, a total of 19,500 cubic yards of sand. At a cost of about $7,500 each, it represents approximately $100,000—less than the cost of one house! The same amount of beach replenishment sand would still be insignificant and temporary.

Figure 6.18 shows an extremely invasive plan for blocking overwash and storm-surge ebb channels (that is, roads) by adding a tre-

6.17 This drawing illustrates essentially the same mitigation plan as in figure 6.16a except it is shown on a larger scale for comparison with figure 6.18.

6.18 A very invasive plan for reducing storm damage on Folly Beach. Small squares represent homes. The plan involves adding large amounts of sand and blocking a number of roads. Some houses have had to sacrifice their driveway access to the plan. In these, shared parking lots (P) provide nearby access to homes. Had this been in place when Hurricane Hugo struck, property damage would have been much reduced.

mendous amount of sand to the island. It is so invasive as to restrict access to some homes. The plan consists of completely blocking Arctic Avenue, the first shore-parallel road, as well as much of Ashley Avenue, the second shore-parallel road. Access to most of these homes would be from parking lots located in previously empty portions of the blocks. All homes to which access is available only from central parking lots are represented by shaded squares. Shaded areas with large *Ps* are the proposed parking areas. The third shore-parallel road is left open, as is the fifth. The fourth and part of the sixth roads are blocked. Direct driveway access to homes in the third, fourth, fifth, and sixth blocks is available from the third and fifth streets back (houses shown by solid squares, driveways by lines). There would, obviously, need to be much redesigning of the driveways. Driveways that previously ran from the fourth street to homes in the third block would be rerouted to run from the third street to the fourth block, and so forth. A few homes on the back side of the island would need to be reached from central parking lots, as on the front side.

The plan shown in figure 6.18 is unquestionably extreme, but money would be better spent, and serve the people better, if it were used to block roads and increase the sand volume of the island rather than to replenish beaches, especially in the case of high-density development of the central portion of Folly Beach. The artificial dunes shown in figure 6.18 represent about 20,000 linear feet (6,100 meters) of the type of dune shown in figure 6.15. The total volume represented is about 150,000 cubic yards (115,000 cubic meters)—a volume that would be a very small beach replenishment project. The portion of the community shown in figures 6.16, 6.17, and 6.18 is between 3rd Avenue east and 7th Avenue east, across the width of the island. This is approximately half of the central part of the community. The total cost would be about $750,000, much less than a beach replenishment project, and the sand is "permanently" emplaced; that is, the sand will still be there when the next major storm occurs, unlike sand placed on the beach, which will start being removed immediately by fairweather processes.

Some obvious costs involved in the plan as presented need further study. For example, the dunes need to be vegetated in order to stabilize them. Vegetation plantings are often done on a volunteer basis, so the cost is variable (see especially Broome, Woodhouse, and Seneca, 1982b). In addition, the cost of re-orientation of driveways could be considerable if they are paved. Preferably, driveways are made out of gravel or left completely unimproved to maintain groundwater recharge, decrease surface runoff, and reduce surge channeling.

Questions may be raised about increased flooding on islands if some of the proposed storm-surge ebb reduction methods are employed. Based on observations made in undeveloped areas of the Yucatán Peninsula of Mexico after Hurricane Gilbert, it appears that flooding would not be a significant problem. In the Yucatán during Gilbert surges were comparable to or higher than those on Folly Island during Hugo. Areas where several rows of well-vegetated dunes existed were absolutely free of any evidence of storm scouring except where a beach access road existed prior to the storm. At that location, a 7-foot (2-meter) channel was scoured by storm-surge ebb flow. The dunes that had not been excavated slowed ebb flow enough to prohibit scouring but did not cause any excess flooding. The point is that the surge water will return in time frames of hours if some road blocking method is employed versus fractions of hours (thus much higher flow velocities) when storm-surge ebb channels are exploited.

It is recommended that less drastic methods than those described above be tested first. Perhaps something as simple as modified "speed bumps" on the shore-perpendicular roads could slow ebb flow enough to reduce scour. Similar features are illustrated for Texas by the Texas General Land Office (1991, p. 18). Perhaps adding some roughness to the road surface by paving with rounded pebbles would help, although increased turbulence might negate the benefits of slightly lowered flow velocity in terms of reducing erosive capability of the flow.

The Sand Commandments

One of the great ironies of developed barrier islands is that everyone is living on a giant pile of sand, but the development has immobilized the pile! All of that sand . . . but communities are desperate for sand to nourish beaches and rebuild dunes—the landforms of sand actively or passively destroyed by the development. Sand is a precious commodity, in short supply on most islands and shorefront communities, and every grain should be safeguarded. Typical soft stabilization measures (as discussed in chapter 5) include beach replenishment, dune building, beach bulldozing, and not much else. We propose additional measures as described below, especially the less traditional methods of adding sand and vegetation to the *interior* of the island.

No matter what soft stabilization measures are utilized, the geologic point of view takes into account the overall sand budget. The island, the beach, the dunes, the tidal deltas (old and new), and offshore features are all linked as one sand system. There are locations within this large sand system where sand is less mobile than others. For example, there is an advantage to adding sand to construct interior dunes over adding sand for traditional beach replenishment. Sand emplaced on the interior of the island will last much longer, at least in a several-decades time frame (i.e., until erosion catches up with it), whereas sand placed on the beach, while offering some protection, is easily removed by fairweather processes as well as storm processes, and so it is temporary, that is, usually remaining less than five years.

In light of the precious nature of sand and the concept of the overall sand system, there are some important Sand Commandments to keep in mind:

(1) Each island and oceanfront community should control its sand resources. No sand should be lost in offshore dumping or hopper dredging. All dredged sand should be dumped somewhere on the island. Not one grain of sand should leave the island!

(2) There should be no robbing Peter to pay Paul. Never should sand be taken from a low-risk part of an island to replenish a dangerous or higher risk part of the island. By so doing, you only end up with two unsafe areas.

(3) Every island should own its own dredge so that when inlets are dredged for navigation, the sand can be put on the island's beaches.

(4) Every island and oceanfront community should have its own mainland sand source, located well inland away from the shoreline and the beach/dune systems of other communities.

(5) Every island community should have a "sand plan" as to where post-storm sand cleanup will be placed (e.g., to repair dune gaps, rebuild T-mounds, add to interior dunes, put back on the beach).

The Sand Commandments could include Vegetation Commandments: Natural vegetation offers such good protection that not one blade of grass, not one shrub or tree should be needlessly cut.

We have so far looked at mitigation approaches on the front side of islands (chapter 5) and, in this chapter, at a more whole-island concept of mitigation strategies on island interiors. There is one important system remaining, however, that often gets the shortest look in terms of planning and management: tidal inlets. In the next chapter we'll take a look at these "in-between" environments.

7 The In-Between Case: Inlets and Property Damage Mitigation

Effective land-use planning and zoning ordinances require an intimate understanding of coastal processes in order to delineate the most hazardous areas and avoid potential hazards. Some of the coastal storm processes that are within the experience of the general public are beach and dune erosion, sand overwash, and storm-surge flooding. Missing from the general body of knowledge, however, is an understanding of the complexity of inlets, their dynamics, and their importance to barrier islands.

Inlets are located between islands, connecting backwaters and ocean. Inlets require special consideration in applying principles of property damage mitigation because a special set of natural processes are at work. Communities and property owners close to existing and potential inlets face a special set of hazards.

Tidal inlets are not fixed features. Inlets can migrate in one direction or another, reverse direction of migration, open and close, narrow or widen, or follow any combination of these patterns. Because of these dynamics, some states (e.g., the Carolinas) designate *inlet hazard zones* or *inlet hazard areas* as special "areas of environmental concern" which are subject to more rigorous management requirements. You might be surprised to learn that the bulk of the sediment volume underlying most barrier islands was deposited in former inlets, indicating just how wide an area is influenced by inlet processes.

Each inlet is unique and should be studied carefully before attempts are made at stabilization with jetties or "improvement" (i.e., dredging). Some examples of problems associated with modern inlet dynamics are examined in the following sections. Table 7.1 summarizes the range of inlet types. Relict and historical inlets are now closed; however, their location and recognition is important because inlets can

7.1 Aerial photograph of South Padre Island, TX, shortly after the passage of Hurricane Allen (1980). The overwash passes are "permanent"; that is, they are occupied in every major storm. The road shown here was cut in at least six places by the hurricane where it crossed these passes. Unlike these in Texas, overwash passes and the sites of new inlets on East Coast barrier islands are often difficult to predict. Photo furnished by Robert Morton.

re-form in previous positions, as happens on Texas islands (fig. 7.1), or they may leave behind a characteristic geomorphic "signal" that may influence whether or not an inlet will reopen in a nearby location, as happens in the Carolinas. Even locations of artificially closed inlets may be re-excavated in the next big storm and the inlet reestablished.

7.2 A generalized diagram of tidal deltas. Each barrier island inlet has an ebb tidal delta (in the ocean) and a flood tidal delta (in the lagoon or sound). As shown by the arrows, much of the volume of the ebb tide follows the main channel, while the flood tide, at least initially, is restricted to flow along the margins of the inlet. The ebb tidal deltas are particularly complex because of the interaction of wave currents, tidal currents, and breaking wave energy. Based on a drawing by Miles Hayes, 1979.

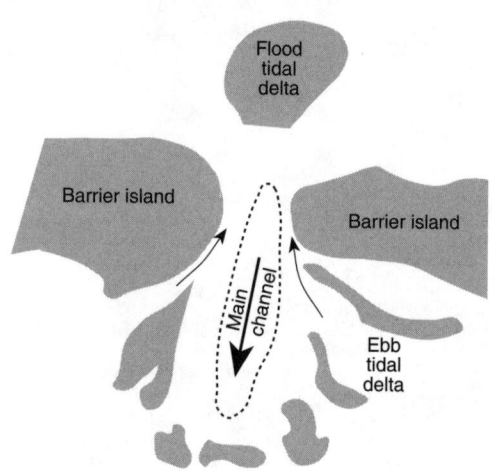

Table 7.1 Classification of Inlet Types on U.S. Barrier Islands

Origin	Age			
	Modern		Historical	Relict
Natural	MIGRATING	STABLE	CLOSED	CLOSED
Artificial	maintained		reopened	reopened
	relocated	closed		
		opened		

Modern inlets, whether natural or artificial, often migrate over considerable distances, eroding not only the adjacent shoreline but erasing the entire downdrift island as new updrift island forms. Even so-called stable inlets, which may not be migrating, often show significant widening in adjustment to increased volumes of flow in storms, and stable inlets may switch to a migratory mode.

For property damage mitigation studies, each type of inlet must be evaluated in terms of past, present, and potential future behavior. In most cases, historical behavior has proven to be a good measure of future behavior. And while the location and habits of modern inlets are very important, relict and historical inlets should also be considered "inlet hazard areas" when evaluating islandwide risk and planning mitigation.

Inlet Dynamics

Inlets form during storms. Most probably form when a storm system passes over an island and the water piled inside the lagoon by landward-directed winds is suddenly forced seaward by the reversed winds. This phenomenon is illustrated in figure 3.12. The water forced across the island catastrophically seeks out low and narrow island stretches to carve new inlets. Sometimes existing inlets can handle the seaward flow of water but are greatly widened during a big storm. When Pamlico Sound water pushed seaward after the

1962 Ash Wednesday winter storm, the half-mile-wide (0.8 kilometers) Oregon Inlet widened to 2 miles (over 3 kilometers) in just a few hours. A number of inlets, especially in Florida, have been created artificially. In 1918 Lake Worth Inlet at the northern end of Palm Beach was cut, probably by the use of mule and drag pan. The lagoon behind is filled with salt water but retains its freshwater lake designation from the pre-inlet days. The inlet at the southern end of Palm Beach (South Lake Worth Inlet) was cut in 1927. Further north, Fort Pierce Inlet was cut by blasting through hard, well-compacted limestone.

Much is known about the behavior of individual inlets, though little of that information seems to be getting to coastal planners and decision makers. Inlets show two common types of behavior that may be categorized as "stable" and "migrating" (see table 7.1). Geologic studies have shown that tidal range is a major factor controlling the types of inlet present and the migratory behavior they exhibit. Whether or not an inlet migrates is a function of longshore drift of beach sand. If there is a strong preferred direction of sand movement on adjacent beaches, the inlet will migrate in the same direction as sand pours into the inlet, forcing it to migrate. Some inlets are grounded in rock and are more or less permanent in their location, regardless of the tidal range or wave direction.

All inlets have associated sand shoals called tidal deltas, formed when sand moving along-shore is interrupted by the flow of water in and out of the inlet. A flood tidal delta in the estuary forms by the flooding or rising tide carrying water and sand into the inlet and the estuary. A corresponding ebb tidal delta forms in the ocean by sediment reworked in the ebbing flow of the tidal waters. The size of the flood and ebb tidal deltas is controlled by tidal range and wave energy. In general, the greater the tidal range, the larger the tidal deltas, and the greater the wave energy, the smaller the ebb tidal delta will be (fig. 7.2).

Within the inlet is a channel that may occupy anywhere from 5 to 50 percent of the actual inlet width. In most inlets the channels

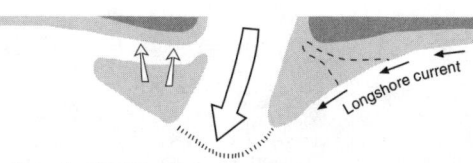

Stage 1 - Initial Growth

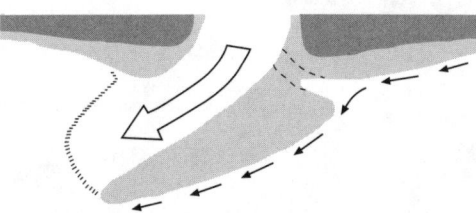

Stage 2 - Extension

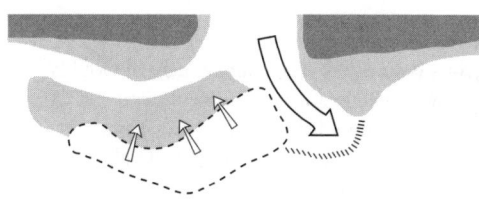

Stage 3 - Abandonment

7.3 Each inlet is unique, but commonly the shoreline behavior of islands on either side of an inlet is determined by the location of the channel (large open arrows) in an inlet. This diagram shows a common sequence of events at an inlet. The sand in the longshore current system (small solid arrows) flows into the channel and forces it to bend as shown in stage 2. A passing storm straightens the channel and causes a large volume of sand to go ashore (small open arrows) on the downdrift island (stage 3). At stage 2, the downdrift or left island is experiencing rapid erosion near the inlet. At stage 3, the updrift island is experiencing relatively rapid erosion while the downdrift island is accreting. This is an important method of transferring sand from one island to the next, but this process stops working when the channel is dredged (see fig. 7.14, New River Inlet) or when jetties are constructed. (After figure by Miles Hayes, 1979.)

7.4 Aerial photograph of the east beach area of St. Simons Island, GA. The channel here in Goulds Inlet is immediately adjacent to the downdrift island, causing accelerated shoreline retreat. Residents have constructed a rock revetment to halt this erosion (see fig. 5.8).

7.5 The outline of the ebb tidal delta can clearly be seen by the line of breaking waves here on the small inlet between Tybee Island and Little Tybee Island, GA. Clearly this tidal delta contains a huge volume of sand.

change location depending on the flow of sand into the inlet (fig. 7.3). The location of the channel may be a factor in erosion of inlet margins. After storms, inlet channels tend to be "straightened out" as the result of large water flows exiting the lagoon during storm-surge ebb. Incoming and outgoing tidal flows tend to follow different paths across the tidal delta, resulting in a complex pattern of bars and small channels on both the ebb and flood deltas (fig. 7.4)

Sand shoals that form part of the ebb tidal deltas where tidal range is large and waves are small may extend for quite a distance on either side of the inlet (fig. 7.5). The ebb tidal deltas off Georgia islands sometimes extend seaward more than 2 miles (over 3 kilometers). The

shoals actually act to protect some of the island shoreline by blocking or refracting wave energy. One result of the shoals' interaction with waves is a characteristic bulge in the island shoreline near the inlet, giving the island end a "drumstick" shape. These bulges of sand form on either side of the inlet as a result of the interaction of ocean currents, tidal inlet flow, and sand supply in the nearshore system. The bulge of sand builds a wide beach, but one that moves as the inlet moves. Once the bulge moves on due to inlet migration, rapid "erosion" again occurs. As the channel migrates within the inlet, the location of the sand bulge changes accordingly.

Some communities obtain sand from ebb tidal deltas for beach replenishment (e.g.,

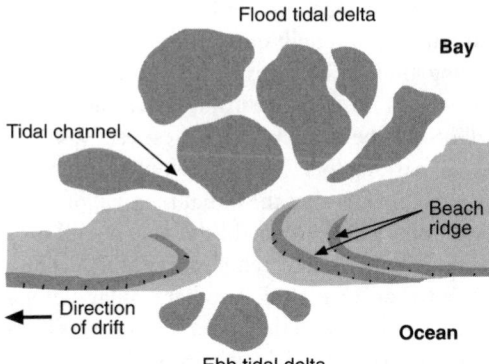

Active inlet features

Flood tidal delta

Bay

Tidal channel

Beach ridge

← Direction of drift

Ocean

Ebb tidal delta

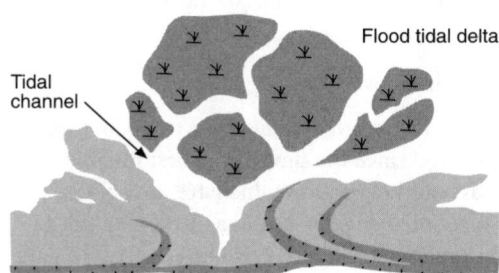

Incorporation of old tidal delta into island after inlet closure

Tidal channel

Flood tidal delta

7.6 Flood tidal deltas, the bodies of sand forced into the lagoon by inlet tidal currents, are often incorporated into the island once the inlet is closed. Widening of barrier islands by the process of flood tidal delta incorporation, as illustrated here, is important in the barrier island migration process. (After Fisher, 1962.)

Hilton Head Island), but this is not a wise move. Changing the shape of the tidal delta will change the nature of the waves striking the island, which will affect patterns and rates of erosion.

Relict Inlets

The term *relict inlet* refers to any inlet that was open in the past but that is older than recorded history and does not appear on any reliable map or chart. Relict inlets are inferred from geomorphic and geologic criteria. An excellent study of relict inlet locations along North Carolina's Outer Banks barrier island chain is provided by Fisher (1962). He investigated the geomorphic expression of former inlets and established the criteria by which relict inlets can be readily identified in the field. Distinctive geomorphic features remain as remnants on the island after the inlet is closed and also can be used along with historical and geographic information (charts, maps, ship's logs) to help locate relict inlets. Common geomorphic criteria include old flood tidal deltas on the back side of an island where no inlet exists (as discussed above, the delta is constructed from sand moved into the lagoon by tidal flushing while the inlet was open), transected by old tidal channels (fig. 7.6). Fisher's study indicated that up to 30 tidal inlets have existed for at least some period of time over the last 400 years along the Outer Banks (fig. 7.7). David Stick, Outer Banks historian, notes that more

than 20 inlets have formed here, been around long enough to be named, and then disappeared. Only three of those—Oregon, Hatteras, and Ocracoke—are presently open and considered important waterways.

Small islands can form in lagoons, landward of the inlet and tidal delta, and are preserved after the inlet closes. These features form as waves roll through the inlet, washing sand and oyster shells up onto salt marsh banks. Such features are a sure sign that the adjacent barrier island area was once an inlet and that it should be examined for evidence of potential new inlet formation. The old relict inlet may have been stable, occupying a fixed position until closed, or it may have migrated. If it migrated, it may have left behind multiple incorporated flood tidal deltas and/or lines of small islands.

The town of Atlantic Beach on Bogue Banks, North Carolina (see fig. 6.2, stops number 1, 2, and 4), illustrates the concept of locating development in a high-risk relict inlet zone (fig. 7.8). The central part of the community is built on the natural fill of a historical tidal inlet that persisted from about the mid-1700s until at least 1800 (Fisher, 1962). Massive dune ridges distinguish the island from the low-elevation historical inlet area. Unfortunately, the area of most intense development had no such protective ridges, and whatever sand dunes were present were excavated for siting of the dense development of Atlantic Beach. A great deal of property is at risk that otherwise would have

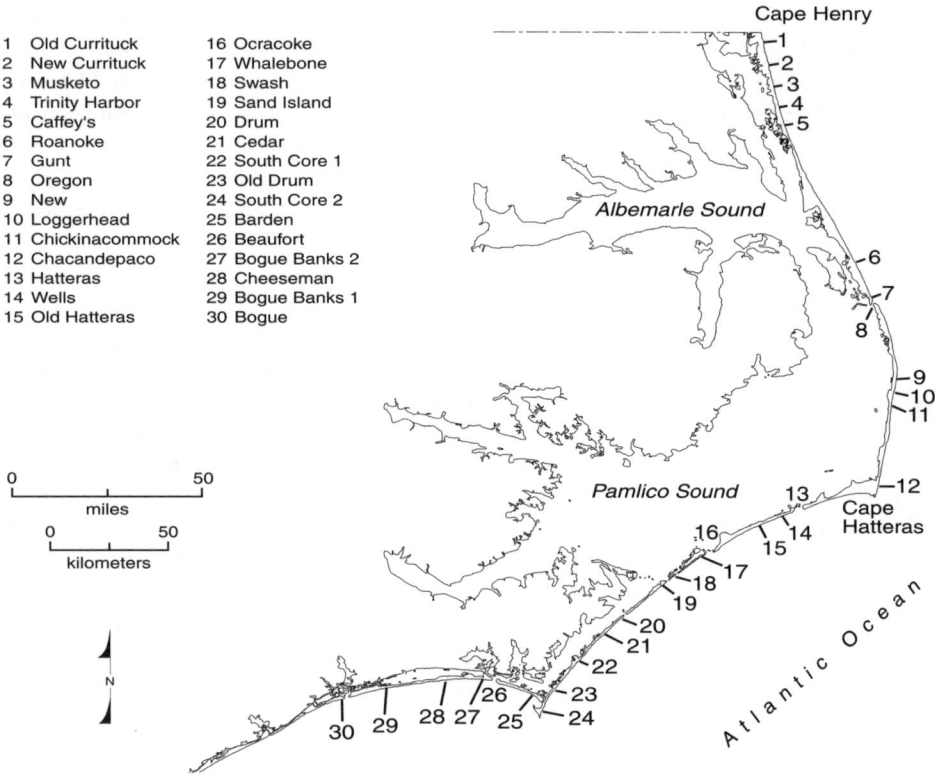

1 Old Currituck	16 Ocracoke
2 New Currituck	17 Whalebone
3 Musketo	18 Swash
4 Trinity Harbor	19 Sand Island
5 Caffey's	20 Drum
6 Roanoke	21 Cedar
7 Gunt	22 South Core 1
8 Oregon	23 Old Drum
9 New	24 South Core 2
10 Loggerhead	25 Barden
11 Chickinacommock	26 Beaufort
12 Chacandepaco	27 Bogue Banks 2
13 Hatteras	28 Cheeseman
14 Wells	29 Bogue Banks 1
15 Old Hatteras	30 Bogue

7.7 The location of historical inlets on the Outer Banks of North Carolina. Each of these inlets was open for a sufficient length of time to be named and put on a chart. Included here (numbers 8, 13, 16, 25, 26, and 30) are present-day inlets. New inlets formed since the 1960s have been artificially filled. Most inlets form by water rushing across the island in a seaward direction. The presence of a large body of water (such as Pamlico Sound) encourages this process. (Modified after Fisher, 1962.)

been safe behind several rows of high dunes. Morehead Avenue, the main road leading to the causeway and the mainland, is located right in this low-elevation, historical inlet location! In a big storm penetration of overwash waters will flood the evacuation route.

Recommendations for this area are to build interior dunes (as discussed in chapter 6) behind the front row of buildings and to build

Morehead Avenue (the road leading to the causeway) up over a rebuilt dune. A drawback to replacing interior dunes, especially for the businesses right on the waterfront, is that the dunes will need to be constructed in these establishments' parking lots. This plan would be unpopular, but consider that the initial removal of sand for the siting of these establishments put all of the real estate located behind them at risk. The rule is to conserve sand. Another similar method of adding volume to the island interior was illustrated on Galveston Island, Texas, after the devastating 1900 hurricane that killed 6,000 people. The island elevation was raised by bringing new sand onto the island, which helped greatly to reduce the risk of flooding (see fig. 6.3).

Historical Inlets

The term *historical inlet* refers to open inlets observed since the time of the first reliable charts and maps, but which are now closed. These inlets may be as old as Roanoke Inlet, through which Sir Francis Drake sailed in June 1586 with supplies for the Roanoke Island Colony and of which only geomorphic remnants now remain, or as young as "New Pawleys Inlet," opened on the southern end of Pawleys Island, South Carolina, by Hurricane Hugo in 1989, which was artificially closed immediately after the storm (see fig. 3.11). Even younger is Little Pikes Inlet, opened across Westhampton Beach, New York, in 1992 and closed in 1993 by the Army Corps of Engineers.

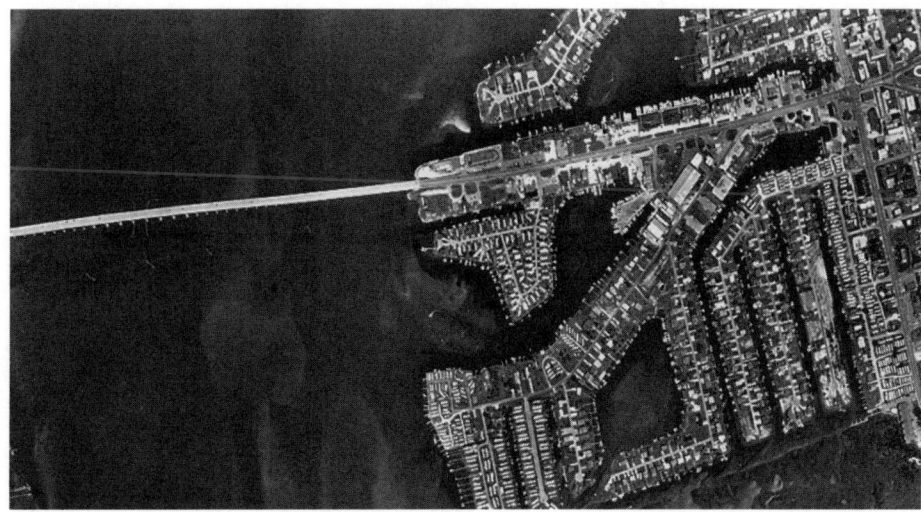

7.8 Aerial view of a dredge and fill area on the sound side of Bogue Banks, NC. Bogue Banks is on the right; a causeway across Bogue Sound connects it to the mainland. Many of the "buildings" shown here are house trailers. This development, started shortly after World War II, was made by dredging salt marsh, a process that is no longer allowed. This marsh was the site of a relict flood tidal delta similar to that depicted in figure 7.6.

Historical inlets often result in modern zones that are highly vulnerable to inlet formation. For example, the narrowest portion of Bogue Banks, North Carolina, is near the center of the island and constitutes a portion of the town of Emerald Isle. Figure 6.2 shows, near the right-hand side of the lower figure (stops number 10 and 11), the location of two inlets opened dur-

ing Hurricane Hazel in 1954. The inlets were filled in by dredging of sand from the lagoon (fig. 7.9). These former inlet locations are easy to spot on the ground where black-stained shells from the lagoonal fill are mixed in with dune sand. Black staining of shells is characteristic of the lagoonal environment and is testimony that sediment was dredged from the lagoon to fill the inlet and raise the island's elevation as the inlet was filled. The island is very narrow in this area and extremely low in places. Although a small flood tidal delta was formed in the lagoon while the inlet was open, the delta's size was limited by the short length of time of the inlet's existence. If the inlets had remained open for a longer period of time, a large tidal delta might have formed in the lagoon, widening the island and actually reduc-

ing the likelihood of inlet formation in future storms. As it is, both former inlet sites are strong candidates for reopening and both are occupied by houses!

In terms of storm damage mitigation, this area of the island should be treated as if it were a present-day inlet and should be designated as an inlet hazard area. Several options exist for any site on a barrier island where inlet formation has occurred historically or can be predicted to occur in the future: (1) add elevation with off-island sand to make it a less likely location for a new inlet (i.e., build an interior dune field); (2) stabilize the margins of the area with vegetated sand ridges as if it were an inlet, giving a degree of predictability of where a new inlet might open; (3) prohibit development in the area; and (4) preserve and even expand back-island marsh areas of the former tidal delta.

7.9 This low, flat portion of Bogue Banks, NC, was the site of an inlet cut by Hurricane Hazel in 1954. It was artificially filled to close it, and buildings have since been built where the channel of the inlet once existed. There is a strong potential here for the development of a new inlet during the next big storm.

Inlet Hazard Area, Historical Inlet:
Folly Island, South Carolina

The pattern of erosion on Folly Island is such
that there is a bend in the island outline near
its northern end. The inflection point marks
the area of an inlet that was open during the
Civil War, separating Folly Island from Little
Folly Island to the north. Figure 7.10 shows
how the islands looked during the Civil War.
The inlet filled in naturally and its former loca-

7.10 Our rendition of a Civil War–era map of Folly
Beach and Morris Island, SC, showing the location
of the Morris Island Lighthouse and Fort Wagner.
The lighthouse is now at sea (figs. 4.16 and 5.19).
The inlet between Little Folly Island and Folly Is-
land is now "The Washout" of Folly Beach. Long
Island and Black Island, behind Folly Island, are
prehistoric barrier islands. (After Stringer-
Robinson, 1989.)

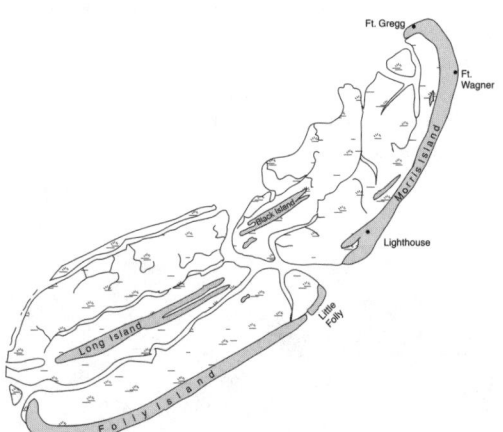

tion remains very low in elevation and is frequently overwashed by storms. Appropriately,
this vulnerable zone is now referred to locally
as The Washout and is said to be the second
most popular surfing location on the East
Coast after Cape Hatteras, North Carolina.

Arctic Avenue, the oceanfront road in the
center of town, is gone, fallen prey to the high
erosion rate in this zone. Ashley Avenue is the
oceanfront road here. Houses and part of
Ashley Avenue were destroyed in another large
washout immediately north of Folly's inflection point. The houses will not be rebuilt because of relatively new South Carolina coastal
regulations. Ashley Avenue has been rebuilt,
however, and is located partially in the salt
marsh because of the narrowness of the island
in this area. A large revetment was emplaced
to protect the road, replacing a smaller revetment that was destroyed by Hugo.

South Carolina seems committed to saving
Ashley Avenue through the washout zone to
allow access to northern portions of the island
for a small number of residents and for a satellite navigation station on Coast Guard property at the very northern end of the island. Relocation of the road is not possible because the
island is so narrow that there is no room for
the road to be moved back. This truly is an
area that should be abandoned. The cost of
maintaining the road will ultimately exceed the
value of the property to which it provides access. The Coast Guard station is not active and
the satellite navigation equipment certainly

could be moved to a safer location. Since the
structures and property are not essential or
could easily be relocated, the economically
wisest option would be to abandon the buildings and leave the area for public recreation.
Real estate interests often prevail over wisdom,
however, and the purchasers of property in
such locations aren't likely to be informed of
how ephemeral the access to their new property may be.

Modern Inlets

Modern inlets are, in a sense, also historical
inlets. The distinction, as we refer to them, is
that these have remained open. In other words,
these are "today's" inlets, the ones we have to
deal with and worry about in terms of their active processes with respect to potential property damage.

A Stable Modern Inlet:
Bogue Inlet, North Carolina

The western end of Bogue Banks, North
Carolina, at Bogue Inlet is a high-hazard zone.
Here the State of North Carolina Division of
Coastal Management has designated an Inlet
Hazard Area, recognizing the dynamic habits
of the inlet. Historically, Bogue Inlet expands
and contracts depending on the channel location within the inlet, but the inlet itself does
not migrate. When the channel is near one
margin or another, sand from the margin

pushed into the channel is swept away. Structures in this low-elevation portion of the island are subject to overwash and flooding. Areas within the known zone of expansion and contraction of stable inlets should be avoided. The study of historical records may establish an approximate set of limits for the inlet hazard zone.

A Migrating Modern Inlet: New Topsail Inlet, North Carolina

From studies of individual inlet behavior, more comprehensive and informed zoning and land-use planning decisions can be made. New Topsail Inlet, the southern boundary of Topsail Island, North Carolina, is a good example of the "individuality" of inlet behavior. New Topsail Inlet historically has migrated consistently to the south. Professor William Cleary of the University of North Carolina-Wilmington has studied the historical habits of North Carolina inlets in great detail, including the effects of inlet management either by jetty stabilization or by dredging. He found that some of the general rules of inlet behavior predict very nicely the behavior of New Topsail Inlet and the subsequent impact of potential property damage as well.

Figure 7.11 is a photo collage of long-term shape changes of New Topsail Inlet. Points labeled 0 and 1 are fixed so that southerly migration of the inlet to the left can be visualized. How the changing shape and orienta-

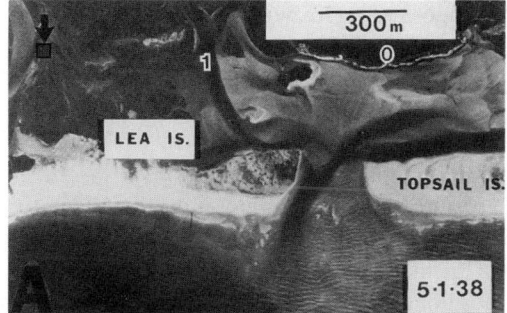

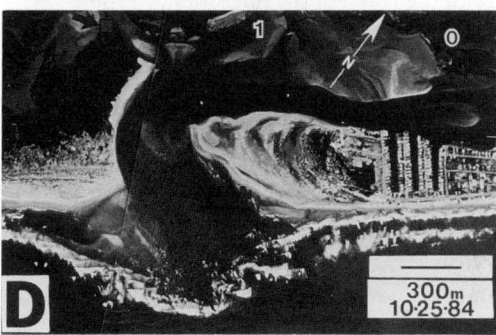

tion of the inlet channel changes the shape of the ebb tidal delta shoals and the corresponding change in shoreline erosion also are shown. In addition, the channel itself can impact the shoreline position on the margin of the inlet and cause significant erosion.

Figure 7.12 shows the history of the southerly migration of New Topsail Inlet. Between 1738, when the inlet opened, and the present, the inlet has migrated 6 miles (9.6 kilometers)! The dates on the figure refer to the year in

7.11 Shape changes in New Topsail Inlet, NC, from 1938 to 1984. The points labeled 0 and 1 are fixed reference points. Note that the bulge on Topsail Island (to the right) moves as the island lengthens and the inlet migrates. Movement of this tidal bulge causes severe shoreline retreat from the standpoint of the owners of fixed structures on the island. Courtesy of William Cleary.

which the inlet was at a given position.

The main point of the figure is that, as the inlet migrates, the drumstick shape of the end of the island also migrates. The fat end of the drumstick, the previously mentioned shoreline bulge, is moving with the inlet and thus is temporary at any one location. As the inlet migrates, the bulge moves, causing shoreline retreat at its previous location.

This natural process of inlet migration shouldn't cause any problem, except when development is unwisely sited within the active inlet zone, especially if located downdrift or "in the way" of the migration. In some states, such as Florida, few inlets remain in an unengineered state, without jetties, so inlet migration is no longer a problem. The migration problem, however, is replaced with a shoreline erosion problem caused by the sand-trapping role of the jetties (see chapter 5).

From Long Island to Florida to Texas, the coast abounds with communities sited next to inlets with histories of migration, and the cost of maintaining such inlets stable (metastable) positions by dredging or through jetties is another major growing expense of the unreliable barrier island infrastructure. Those who are prudent in regard to their property investments will demand studies to define inlet hazard zones for land-use planning. Better yet, prudent property owners will stay away from inlets. Figure 7.13 shows diagrammatically the differences in managing coastal zoning for migrating versus stable inlets.

7.12 Diagram showing the migration history of New Topsail Inlet, NC. The inlet opened in 1738 and has been migrating ever since. There is a strong likelihood that a new inlet will open some day and the whole process will begin again. Courtesy of William Cleary.

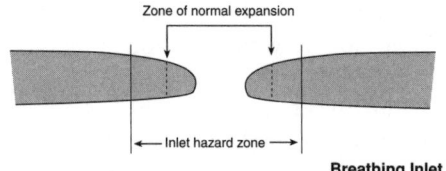

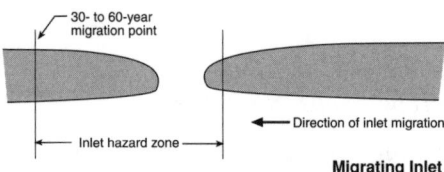

7.13 Diagram showing the designation of inlet hazard zones for migrating inlets and breathing (non-migrating) inlets.

Migrating inlets can have a profound effect on property damage mitigation. The interior of Topsail Island is open and flat, a topography typical of younger barrier islands built in the track of a migrating inlet with insufficient time to form large dunes. Potential for new inlet formation exists at all narrow, low-elevation locations along this island, especially where finger canals are present. The frontal dune, where it exists, protects the interior "grasslands" and maritime forest.

Topsail's opportunities for maintaining elevation naturally by overwash and dune formation have been altered through development because the frontal dune is kept fixed, not al-

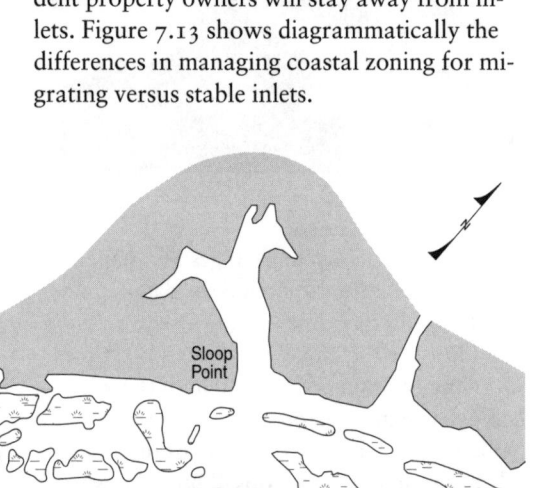

lowing sand overwash and dune migration to move sand in a landward direction. Thus, the island interior remains at low elevation and narrow. This problem is common on most developed islands. The elevation of Topsail (and other islands) could be artificially maintained through a program of replenishment (or augmentation) of sand volume in the island's interior. Several methods by which this might be accomplished include sand fencing, bringing in new sand and building and stabilizing interior dunes with vegetation, and establishing maritime forests.

A Dredged Modern Inlet: New River Inlet, North Carolina

Some modern inlets are artificially stabilized or held in place through dredging. The inlet channel of New River Inlet, North Carolina, is artificially held in place by channel dredging which, in turn, determines the position of inlet-associated shoals, which provide protection for the town of North Topsail Shores on Topsail Island (fig. 7.14). As maintained, the inlet channel makes several sharp turns between ocean and lagoon. Left to its own devices, the channel would create a more direct, less winding link, and the inlet would migrate to the south as it has done historically. A storm is likely to realign the inlet through a narrow section of West Onslow Beach or at least to shift the inlet channel to the south. Once the inlet shifts, rapid erosion will occur

on the northern end of Topsail, as it will no longer be in the "lee" of the inlet shoals.

The dynamics of New River Inlet are illustrated in figure 7.14, courtesy of Professor William Cleary. The figure shows photographs from 1938 and 1986, before and after the Army Corps of Engineers began maintaining the channel by dredging. White triangles on either side of the inlet in each photograph provide static reference points. Note the change in the size of the ebb tidal delta (the offshore sand shoals delineated by the breaking waves), the change in the width of the inlet channel, and the relative change in offset of the two island shorelines.

There are ramifications of maintaining the New River Inlet channel by dredging: (1) the beach on Onslow Beach (part of the Marine Corps' Camp Lejune) to the north (right in the photos) is now severely eroding and steepening, making it difficult for beach-landing military vehicles and exercises; and (2) the northern end of Topsail Island is out of equilibrium as the inlet wants to migrate to the south, presenting a real danger from flooding and inlet channel-switching that will occur during the next major storm. Development across the width of the island in this zone of potential inlet migration or reformation is at extreme risk.

Figure 7.15 suggests possible mitigation for the very dangerous area adjacent to a migrating inlet. Obviously, the best way to reduce the potential for property damage is

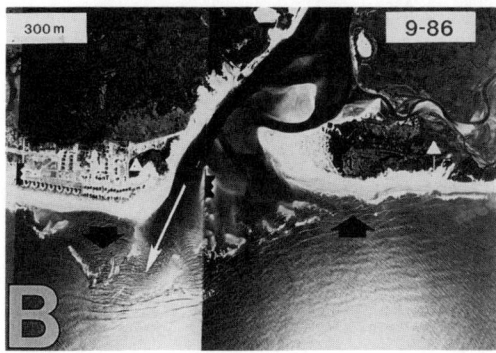

7.14 New River Inlet at the north end of Topsail Island, NC: (a) 1938 (b) 1986. The white triangles on either side of the inlet in each photograph are fixed reference points. Onslow Beach, the island to the right, has retreated 600 feet (180 meters), directly due to the stabilization of the inlet channel by the Corps of Engineers. When a big storm reorients the inlet channel, North Topsail Shores (the community to the left) will suffer severe erosion. Courtesy of William Cleary.

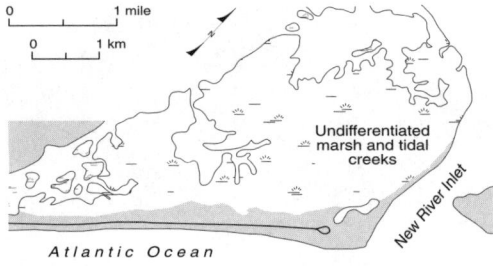

0 — 1 mile

0 — 1 km

Undifferentiated marsh and tidal creeks

New River Inlet

Atlantic Ocean

Problems
Inlet hazard area includes everything within a mile of inlet
Large structures close to beach: overdevelopment
Road washouts make evacuation difficult
Shift in inlet will cause high erosion rates

Assets
Beach presently accreting due to ebb channel position
Extensive backbarrier marsh

Recommendations
Do not rebuild following storm
No new construction seaward of road
Establish frontal dune

7.15 Suggestions for mitigation of the extreme risks to property damage at the north end of Topsail Island adjacent to New River Inlet (see fig. 7.14).

to remove the houses in this zone. Other than that, adding some interior island elevation or well-vegetated sand dunes along the inlet margins may slightly reduce the potential of inlet channel migration during a storm.

A Relocated Modern Inlet:
Sunset Beach, North Carolina

Sunset Beach is the southernmost developed barrier island in North Carolina and actually trends almost east to west. The island is about 3 miles (5 kilometers) long and a little under a half mile (less than 1 kilometer) wide. Sunset Beach has elevations between 10 and 15 feet (3 to 4.5 meters) but only light vegetation in its central developed portions (i.e., few trees or large shrubs). What it does have along most of its length is a protective, wide, vegetated dune field. The central portion of Sunset Beach, between the sand spits at the east and west ends, has historically been accretional. In fact, almost everything (beach and shrub thicket) seaward of the development has built out in the last 30 to 40 years. This accretion zone is a buffer that was significantly reduced in size during Hurricane Hugo but remained intact and protected the houses behind it. The dune field should be encouraged to continue growing and stabilizing for maximum protection.

The presence of such a natural buffer allows the area west of the finger canals to be upgraded to high rather than extreme risk. Storm-surge flooding from the tidal marsh and sound is still a significant threat. Everything east of the finger canals is in an extreme-risk zone. Much of this land is actually artificial fill built on old marsh. Tubbs Inlet has a history of migrating to the west. In 1966 the inlet was relocated at public expense by the U.S. Army Corps of Engineers to the east, approximately

to its 1930 location (fig. 7.16). Between 1930 and 1966 it had migrated all the way to the position of the last finger canal, a distance of approximately one-half mile (almost 1 kilometer). But the inlet will likely migrate again. Unfortunately, when the inlet was artificially relocated, the "new" land (formed by pumping sand) east of the former inlet position was reopened or reprivatized, providing a development rush into an extreme hazard zone. Also, the wide dune field on the western end of the island narrows significantly to the east of the pier. Such relocation projects should be viewed by communities as an opportunity to build in mitigation to reduce risk potential. Instead of permitting new buildings on the newly formed low-elevation sand flat, reconstruction of interior dunes and the establishment of vegetative cover could have been set as prerequisites for development. Better yet, no new development should have been allowed and the state or local government unit should have held title to land more suitable for public recreation, thus avoiding investment vulnerability and the associated extreme to high risk.

7.16 This aerial photograph of Sunset Beach, NC, clearly shows the portion of the island that was reformed by dredging and relocation of Tubbs Inlet a decade before this photograph was taken in 1977. Fortunately for the property owners on this now totally developed artificial spit, Sunset Beach is the only island in North Carolina that is accreting seaward.

Potential Inlet Locations and Newly Formed Inlets

The lesson of Hurricane Hugo was clear: inlets can, *and will*, form across low-elevation, narrow portions of islands during storms. Hurricane Hugo left Pawleys Island, South Carolina, with extensive property damage, overwash, and a new inlet (see fig. 3.11). Northeasters can form inlets, too. The 1962 Ash Wednesday storm formed new inlets across islands in New Jersey as well as at Buxton, North Carolina. All these inlets were filled in artificially, as have virtually all new inlets that have formed across developed barrier islands in the 1980s and 1990s.

Examination of Pawleys Island reveals a spectrum of property damage that appears to have been a function of the presence of protective dunes, setback, and vegetation. Relatively little damage occurred where houses were well elevated, well back from the beach, behind the frontal dune, and enveloped by dune and maritime forest. The most severe damage was found where the interior dunes and maritime forest had been removed for roads, houses, driveways, and parking areas.

The low-elevation narrow spit on the southern end of Pawleys Island was, predictably, the site of an inlet formed by Hurricane Hugo (see fig. 3.11). The new inlet cut off access to a few homes so it was filled in artificially, but plugging it up took two tries.

Inlets on Padre Island, Texas, tend to be temporary. They occupy well-defined passes that

become dry land again shortly after the storm goes by. One of the passes near South Padre is currently occupied by a condominium.

New inlet formation was likely on Pawleys Island in such a low and narrow location and should have come as no surprise (table 7.2). Historically, inlets had previously formed in the same locations and then migrated to the south. Regrettably, houses were rebuilt on precisely the spot where the inlet opened. Mother Nature is giving us a hint here: Avoid this entire spit! A new inlet will open here again and the only viable mitigation strategy is to stay clear.

A fundamental rule is: *Avoid both active and potential inlet zones.*

Table 7.2 Rating of Factors Responsible for New Inlet Formation, Outer Banks, North Carolina

Parameter	Average	Dangerous	Hazardous	Extremely Hazardous
Primary:				
Island Width (feet)	2,300	> 2,600	2,000–2,600	500–2,000
Maximum Elevation (feet)	*	> 12.5	7.5–12.5	< 7.5
Secondary:				
Lagoon Width (miles)	22.4	0–14	14–30**	> 30
Canal Approach*** (feet)	1,500	> 1,500	1,000–1,500	500–1,000
Canal Width (feet)	150	< 100	100–200	200–300
Erosion Rate (feet/year)	4.5	< 3.0	3.0–6.0	> 6.0

* Average 25-year storm surge is 7.5 feet above mean sea level.
  Average 50-year storm wave height is 12.5 feet above mean sea level.
 **Includes sites with lagoon widths > 30 miles in one direction.
 ***Distance between the ocean shore and the farthest canal penetration.
Canal refers to finger canal. The ranking of average to extremely hazardous refers to the relative tendency for an inlet to form. These numbers are designed for the Outer Banks only, but a similar ranking of parameters could be made for any barrier island chain.
Source: Lynch and Benton, 1985.

In the best of all worlds, development would be based on providing maximum aesthetic enjoyment to the property owner, minimizing any risk from natural hazards, and providing a reasonable financial return to the developer. Natural processes and natural settings would be evaluated for their hazard potential. Mitigating future property damage would be a routine part of planning for development (fig. 8.1).

Rules and Reality

Principles and rules often evolve from the trial and error of experience. Realizing that hindsight is a luxury that early developments did not have, some principles still emerge.

The best time for mitigation is at the beginning of development. The initial stage of development in the coastal zone is one of opportunity to minimize vulnerability to natural hazards. Strict requirements can be set on siting (e.g., elevation requirement, rules minimiz-

ing vegetation removal or dune disruption), and the development can be platted to conform with the environment rather than to a grid superimposed on the natural topography and habitats (e.g., roads can curve around or over dunes without cuts, evacuation routes can be sited in the least vulnerable locations, potential overwash passes or inlet locations can

be avoided, lot depth can be adequately established for future relocation). The initial development plan, plat restrictions, and community regulations can set the course for minimizing risk. This principle is offset by the fact that *the majority of coastal communities have grown haphazardly without long-term planning* (development often proceeded from low-risk sites

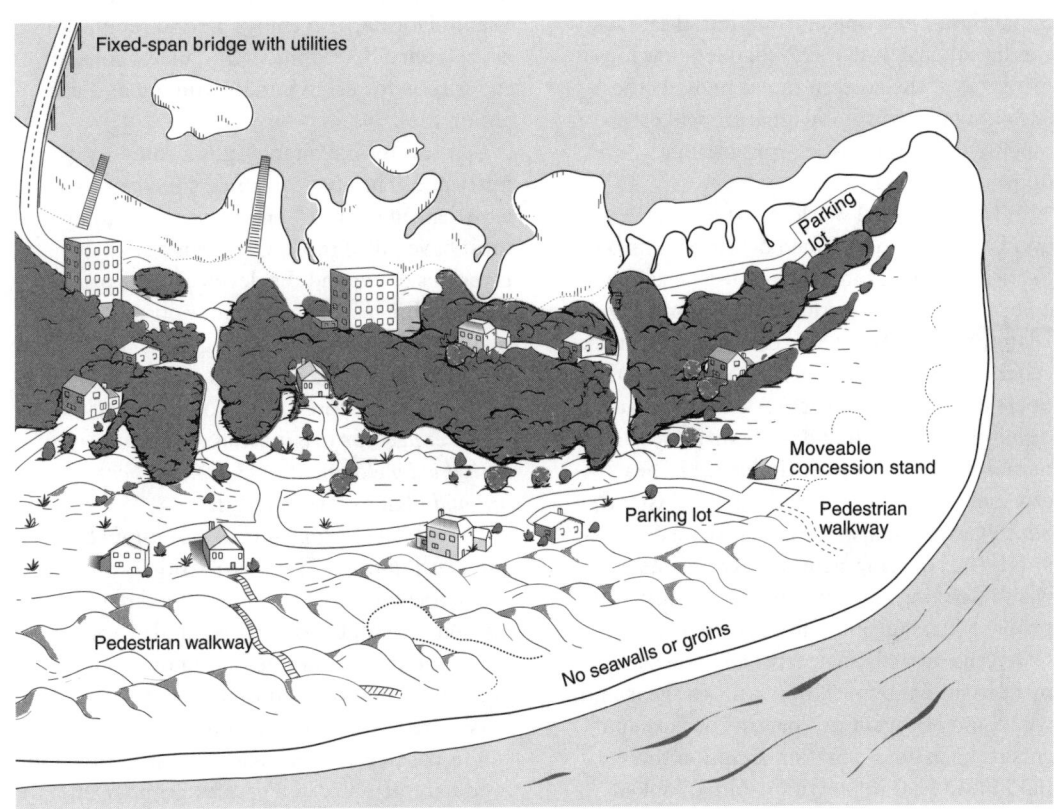

8.1 Pandora's Fantasy. The "ideal" developed barrier island. If you are imprudent enough to want to develop a barrier island, this is the way it should be done: very little alteration of topography or vegetation; highest density of population and all high-rises on the lagoon side of islands; roads curved and going around or over, not through, dunes; access to the beach by walkway only; and parking lots and parks in areas of potential overwash or erosion.

to high-risk sites). In other words, for most coastal communities it's late in the game.

Planned development has the potential to be less vulnerable (at lower risk) than haphazard development. Planning involves establishing island dynamics, risk evaluation, and mapping and working to incorporate natural processes and property development in such a way as to reduce both short-term and long-term hazard impacts. For example, distant initial setback can be sold as "low-risk," planned development, but if the erosion rate is high, or the sites' elevations low, then nature will catch up quickly in the form of storm surge and flooding.

In truth, the real world provides very few good examples of planned developments on barrier islands, primarily because developers/communities do *not* stick to the plan. Kiawah Island, South Carolina, is often cited as an excellent planned development, but several developers later noted some cracks in the "plan" are developing as housing density increases and nature begins to show its influence in beach erosion. The interior of Hilton Head Island, South Carolina, also initially provided an example, but development density now exceeds island carrying capacity and the potential for property damage in a future storm is huge. Given the new planned retirement development to add another 8,600 units to the island, we would not want our parents or grandparents to be in an evacuation situation there! Bald Head Island, North Carolina, looked

good (moderate density development, no cars allowed), but here, too, the seaward incursion of development, disruption of dunes and vegetation, and landward incursion of the shoreline do not suggest long-term low risk. Little Cumberland Island, Georgia, may be a good example (low-density development, elevation, vegetation cover), but its exclusiveness suggests only a few can enjoy this ideal. Sanibel Island, Florida, also enjoys a good reputation as a planned development and meets some of the criteria for using initial planning as a mitigation tool.

Our test for risk mapping is a category 3 hurricane. The above examples have not yet experienced the head-on test of nature, but most have failed the longer term test of the pressures to expand the development, protect the first threatened shorefront house with a seawall, cut new streets through the interior dunes, or locate the new strip mall in that prime maritime forest. A corollary principle is: *Don't expect the original developer to stay with the development, and don't expect the original plan to stay in effect.*

Planned development does not *imply low-risk development.* This observation grows out of the previous discussion of a good development gone bad. Unfortunately it also applies to many young communities along our coasts. Previously undeveloped islands were platted to maximize lot and housing unit sales. Finger canals destroyed back-island marsh protection (and habitat), placed a greater number of

houses in the flood zone, and sometimes raised the potential for new inlet formation, all for the sake of recreational boat access.

Marco Island, Florida, is an example of how access spurs development in high-risk zones, creating greater and greater probability of property loss. The island remained near its natural state until the mid-1960s. Most of the island is less than 10 feet (3 meters) in elevation; hurricanes inundated the island on three occasions in the twentieth century prior to development. The storm of October 18, 1910, raised flood waters 8 to 10 feet (2.4 to 3 meters) above sea level, and in 1947 the flood level reached 7.4 feet (2.3 meters). Hurricane Donna (1960) drowned the island in a 10.2-foot (3.1-meter) flood. This prior history of flooding was ignored by the early developers, and in 1964 the island was subdivided, and airport constructed to bring in prospective buyers, and an extensive system of finger canals was dredged, destroying salt marsh and the attractive habitat of the area. The first access bridge was constructed in 1970 and a second bridge in 1978. Rapid development followed, and by 1980 there were 1,000 houses and 2,500 condominium apartments supporting a population of between 7,000 and 10,000 people. By 1990 those numbers had more than doubled.

The evacuation maps for the Marco Island area tell the story: a very high concentration of high-valued property in a flood zone that includes almost the entire island as well as an extensive area of adjacent mainland. The only evacuation route will flood well in advance of a hurricane landfall, as will many properties. Wind damage will be excessive, and the island already has experienced an "erosion problem" along the shoreline. Like the other islands of southwest Florida, the developmental property carrying capacity of the island was exceeded early in its development history. Nature's clock is counting down for Marco Island's next Donna.

Communities that were poorly planned, that grew haphazardly, or where risk potential has increased can change this trend through islandwide mitigation strategies that preserve, augment, and repair natural environments. Implementation of this recommendation involves taking active steps to repair damage to the natural setting brought about in the name of development. Such repairs will prepare individual buildings, as well as entire communities, for storms, will reduce property damage during storms, and will speed recovery after storms. Thus, coastal residents will be working *with* rather than *against* the natural setting.

When to Mitigate

Property damage mitigation techniques can be implemented on a pre-storm, immediate pre-storm, and post-storm basis. In one sense, a "pre-storm" situation always exists. That is, recovery from any storm should be done with the next storm in mind. Storm recovery should

involve active steps to repair the island itself and to enhance the protective characteristics and capabilities of the natural setting.

Pre-storm mitigation activities include conscientious location of development, elevation and orientation of new roads, changing road orientation or elevation of existing roads, building and vegetating dunes, and updating and enforcing erosion setbacks. Although relocating structures and roadways out of hazard zones is often considered a post-storm activity, it is also possible to consider such actions as important pre-storm mitigation options. Immediate pre-storm activities include sandbagging and removing mobile structures and objects, if possible. Post-storm activities other than reconstruction are implementation of property protection measures, such as relocating structures and roadways out of hazard zones, rebuilding or upgrading structures, restricting density of new development, enforcing new setbacks for relocated and replaced structures, and being aware of and avoiding overwash passes and new inlets. The predicted acceleration of the rate of sea-level rise and increase in frequency and intensity of storms should also be considerations in comprehensive shoreline management policies.

Mitigation Options Summary

Several mitigation options were presented in chapters 5, 6, and 7, with emphasis on nonstructural methods. Mitigation options

were listed and categorized in tables 5.1 and 6.1.

Table 8.1 shows causal relationships between storm effects (table 3.1) and mitigation techniques (tables 5.1 and 6.1). The table indicates that certain mitigation options and storm effects can be causally related. That is, implementation of a given mitigation option will reduce (+) or increase (-) a given negative storm effect. Some options (marked ±) may have positive effects in some places and negative effects in others. An example is a sediment-trapping jetty which, by interrupting the natural alongshore sand transport system, will cause building out of the beach on the updrift side (potentially leading to wider beach and dunes and less potential for property damage) and erosion of sand on the downdrift side. No attempt is made to quantify the effect of each mitigation option in the table. Table 8.1 also shows (along the bottom two rows) whether the environmental impact of each option will be positive (+) or negative (-) (i.e., whether the activity upgrades or degrades the natural environment) and whether the long-term cost will be relatively high (H) or low (L) in general terms. For example, beach replenishment has a high long-term cost because it must be repeated again and again. Replacing interior dunes, however, is low cost because replenishment may be a one-time action.

Mitigation options work to reduce the risk of property damage in different ways. Recall from chapter 1 that a *risk* (in this case, of property being damaged by storms) involves both

Table 8.1   Options to Mitigate Storm Effects, with Long-Term Cost and Environmental Impact

Storm Effects	Mitigation Options																				
	BR	BB	FD	DG	IE	ID	SW	GR	JT	BW	RF	SD	PM	RH	CE	AR	EH	RR	SB	SE	DN
Wind attack											+			+		+				−	+
Flying debris											+			+		+				−	
Direct wave attack	+	+	+	±	+	+	+		±	+	+	+		+	+	+	+	+	+	+	
Oceanside erosion	+	+	±			±	+	+	±							+					
Soundside erosion						+	±	+	±				+			+					
Oceanside flooding	+	+	+	+	+	+	+	±	±	+	+	+		+	+	+	+	+	+		+
Soundside flooding	+	+	+	+	+	+	+	±	±	+	+	+		+	+	+	+	+	+		+
Floating debris			+	+	+						+		+	+	+	+			+	+	
Overwash	+	+	+	±	+	+	+	±	±	+	+	+		+	+	+	+	+	+		
Inlet change		+							±												
New inlet formation			+	+	+				±				+			+					
Sand supply change	+		+	±	+	+	±	+	±	+	+	+				±					
Environmental Impact	+	−	+	+	−	+	−	−	−	−	+	+	+		+	+	+	+		−	+
Long-Term Cost	H	L	L	L	H	L	H	H	H	H	L	L	L	L	L	L	L	L	L	L	L

+=Reduce property damage potential or positive environmental effect.
−=Reduce property damage potential or detrimental environmental effect.
H=Relatively high long-term cost.
L=Relatively low long-term cost.

Key to mitigation options:
BR=Beach replenishment
BB=Beach bulldozing
FD=Raise frontal dune elevation
DG=Plug dune gaps
IE=Raise island elevation
ID=Replace interior dunes
SW=Seawalls
GR=Groins
JT=Jetties
BW=Breakwaters
RF=Replace forest
SD=Vegetate/stabilize dunes
PM=Plant marsh
RH=Retrofit homes
CE=Curve and elevate roads
AR=Active relocation
EH=Elevate homes
RR=Replace roads with dunes
SB=Setbacks
SE=Site elevation
DN=Do nothing

hazards (the physical processes) and *vulnerability* ("improved" property, that is, buildings, utilities, infrastructure). Storm processes cannot be altered, but it is possible to influence or control how intensely the storm's physical processes are *felt* in a given location.

Take a seawall, for example. If you could stand in front of a seawall during a storm, you would certainly see that the wall has no effect on reducing the "strength" of the storm. Immediately behind the wall, however, there would be a noticeable decrease in wave energy *for that location*. Similarly, building interior dunes, thus raising island elevation locally, will do nothing to limit absolute storm-surge elevation, but it will limit the storm-surge inundation *for that location*. These types of approaches are essentially reducing the hazard's impact by modifying the physical processes for small areas of the coast.

Property damage mitigation techniques also may be used to reduce the vulnerability of a building to damage by modifying the building itself, for example, strengthening the building (to resist damage and to limit damage to nearby buildings by wind-borne or water-borne debris) or moving it out of a hazardous zone.

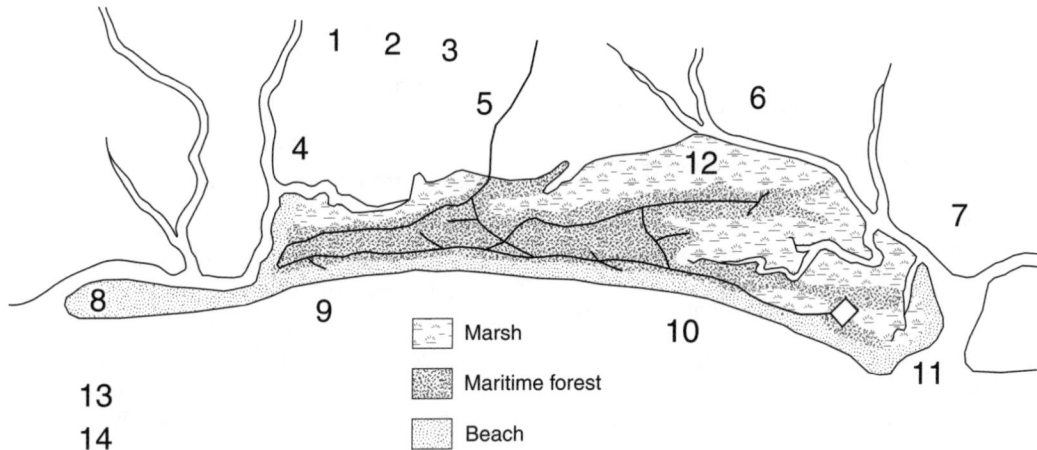

Marsh
Maritime forest
Beach

8.2 Fantasy Island. Another view of the ideally developed barrier island. Numbers on diagram correspond to table 8.2. (Adapted from Pilkey and Neal, 1980.)

Fantasy Island

The best of all worlds is represented in an "ideal" development (fig. 8.2) and outlined in table 8.2, developed along the lines of the Pandora's Island concept in figure 8.1. Given that such initially planned developments are the exception, table 8.2 lists corresponding mitigation actions for existing communities ("Return to Fantasy Island," table 8.2).

The mitigation recommendations presented in this book are all working and well illustrated in various locations along the southeastern U.S. coastline. From the site-specific nature of the study a list of general Principles of Property Damage Mitigation, applicable everywhere, is offered:

Soft stabilization involves some method of augmenting the sand volume of a community or island.
—Sand can be added to the beach, to the dunes, or to the island itself. Sand added to the island is more permanent than sand added to the beach.
—Observe the Sand Commandments and preserve all sand.
—Beach bulldozing may work, but the range of its applicability is limited.
—Sand fencing really only works well where there is both an adequate supply of sand moving onshore and space in front of buildings in which the sand dunes can grow.
—All gaps in dunes that were excavated for building location, ocean view, or ocean access should be plugged.
—Building (or rebuilding) interior dunes adds sand volume to the island and may reduce the effects of overwash and storm-surge ebb.

Building relocation is a sure way to avoid damage: Move the structure out of a hazardous location.
—Many houses are now threatened and can be relocated relatively cheaply.
—Larger structures can be relocated if the economic situation is advantageous.
—Adopt the 10/100-year relocation concept for communities with large structures.

Hard stabilization is not generally recommended because of its cost and long-term deleterious impacts on the beach.
—Shoreline armoring ultimately results in loss of the recreational (and protective) beach.
—The impacts of shoreline engineering can only be fully understood and evaluated after a several-decades-long observation period.

Vegetation, especially maritime forest, is critical for reducing damage from storms.
—Remove as little vegetation as possible, especially forest.

Table 8.2 Fantasy Island: Characteristics of Ideally Developed Barrier Island and Methods to Return to Ideal for Developed Islands

Fantasy Island (undeveloped)	Return to Fantasy Island (already developed)
1. Perpendicular access of utilities; buried.	1. Existing utilities made as safe as possible. Consider burying.
2. Large-craft marina on mainland.	2. No new large-craft marinas.
3. No finger canals.	3. No new finger canals. (Consider filling some?)
4. Back-side erosion control by encouraging vegetation.	4. No additional bulkheading of lagoon side.
5. Fixed span bridge.	5. Upgrade bridges for evacuation; no drawbridges.
6. No disturbance of salt marsh.	6. No dredge and fill. Monitor pollution.
7. No utilities across inlets.	7. Relocate utilities.
8. No development on spits.	8. Relocate development off spits.
9. Minimum disruption of topography and vegetation.	9. Strict planning. Restore frontal, interior dunes, revegetate dunes and forest.
10. No shoreline stabilization except by replenishment.	10. To protect beaches, adopt long-term plan to remove seawalls, as South Carolina has done.
11. No development in inlet hazard areas. Each inlet area studied for historical trends.	11. Rezoning and relocation of development out of inlet hazard areas, including historical, present, and potential inlets.
12. High-rise development on lagoon side of island only.	12. New high-rises on lagoon side only.
13. Development density based on island's natural carrying capacity.	13. Slow or stop development.
14. All buildings exceed codes and regulations (e.g., extra elevation, extra connectors, design to minimize wind impact).	14. Building construction upgraded to reduce hazard impact (e.g., addition of connectors, opening ground-level obstructions to flow, flood proofing).

Note: Numbers correspond to numbers on figure 8.2.

—Encourage marsh or mangrove vegetation growth as a way to slow erosion on the lagoon side of barrier islands.
—Protect and encourage dune vegetation as a way to stabilize sand to offer protection from storm surge and overwash.

Modification of development and infrastructure are ways to change what has already been built to make it less likely to be damaged or to decrease damage.
—Many roads were built perpendicular to the shoreline, making ideal channels to be exploited by storm overwash and storm-surge ebb flows. These roads should be elevated, curved around or built over protective landforms, or resurfaced to impede such flows.
—Many dunes were notched so that roads could go through the dune instead of over it. Here, dunes should be rebuilt and the road rerouted around the rebuilt dune.
—Many homes were built before building codes or when enforcement was poor. These should be retrofitted.
—Breakaway ground-level construction in homes elevated on stilts becomes rubble that rams other structures during storms. All such ground-level construction should be prohibited.

Zoning with nature means identifying hazard areas and avoiding them by proper planning.

—Recognize hazard areas and avoid them.

—Understand inlet dynamics: each inlet is different.

—Create reasonable setbacks based on erosion rates and predictions of erosion rates.

A Complex Picture

Property damage mitigation is at once a simple and a complex task: the principles are simple, but the implementation is complex. Part of the complexity arises because property damage mitigation involves expertise from engineering, the physical sciences, and the social sciences. Engineering, and to a large degree the physical sciences, are relatively straightforward, or at least can be dealt with independently. Engineering involves such things as design of shoreline stabilization structures, beach replenishment, building codes, infrastructure design, and relocation. The applied physical sciences provide information about coastal geology, ecology, biology, and environmental science, and meteorology and climatology—predicting hurricanes and other storms. The social sciences have the truly onerous tasks involved in property damage mitigation. They must address complex issues such as land-use planning, emergency preparedness and emergency response, coastal zone management, and politics. Thus, the social aspects constitute an area

of much needed focus. Coastal geologists can point out inlet hazard areas and likely overwash zones, calculate erosion rates, and so forth. However, it takes a more integrated effort in order for people to understand and accept that we must live with nature, not try to fight it.

The following case studies illustrate that each island/community is unique, and, historically, the approaches to mitigating natural hazard impacts are as diverse as community social structure and politics.

9 Selected Risk Assessment and Property Damage Mitigation Recommendations: Atlantic and Gulf Coasts

Each barrier island and coastal community presents unique challenges with respect to assessing hazards and evaluating vulnerability and risks for coastal property damage (tables 9.1 and 9.2). Variations along the Atlantic and Gulf coasts in geologic setting (e.g., sediment supply and type), climate (e.g., temperature range, rainfall, and type, frequency, and intensity of storms), and oceanographic setting (e.g., tidal range and average wave energy) translate into a wide variety of barrier island morphologies, vegetation cover, shoreline types, and hazards. Added to the complexity of these natural variations is a wide range of development types and densities. Communities ranging from light developments of a few single-family dwellings to dense high-rise condominium and hotel rows require a like range of management and property damage mitigation strategies.

Based on our experience and case studies in hazard assessment, risk mapping, and mitigation recommendations, we have presented in the previous chapters a set of *general* or universal principles of property damage mitigation. General principles serve only as a guide, and each location must be evaluated on its own merit in order to make property damage mitigation recommendations that are site specific (table 9.2). These principles were illustrated and their application implied in specific examples drawn from the variety of areas hard hit by Hurricane Hugo (e.g., South Carolina's Charleston area islands, particularly Folly

Table 9.1   Property Damage Risk Attributes of Islands in the Carolinas

Bogue Banks, NC
Atlantic Beach, Pine Knoll Shores
—moderate elevation, moderate to wide width
—wide range of development settings
—"free" replenished beach
Indian Beach, Salter Path, eastern Emerald Isle
—low elevation, narrow
—historical inlet location
—wide range of development settings
Central Emerald Isle
—very high elevation, very wide
Western Emerald Isle
—wide, moderate elevation
—inlet hazard area

Topsail Island, NC
North Topsail Beach
—inlet hazard area
—low elevation, narrow island
—high- and low-rise commercial development
Surf City
—low elevation
—narrow
West Onslow Beach
—inlet hazard area
—low elevation, narrow

Southern Brunswick County, NC
Ocean Isle Beach
—low elevation
—finger canals
Sunset Beach
—high-density single-family homes
—accreting beach

Grand Strand Area, SC
Entire area
—mainland beach
—high elevations
—very high-density high-rises
—high-density single-family homes in places
—North Myrtle Beach is low elevation

Litchfield, SC
—very low elevation
—artificial dune destroyed by Hugo

Pawleys Island, SC
Southern end
—very low elevation spit
—moderate-density single-family homes
Central
—moderate elevation and well forested
—high-density single-family homes
Northern end
—low and narrow
—ocean to lagoon streets

McClellanville, SC
—not oceanfront
—sits back behind six miles of low marsh
—very low elevation

Isle of Palms, SC
—high-density single-family homes
—benefits from being updrift of jetty

Sullivans Island, SC
—high-density single-family homes
—benefits from being updrift of jetty

Table 9.1 (continued)

Folly Island, SC
Southern end
—low elevation spit
Central
—high-density single-family homes
Northern end
—historical inlet location
—extremely low elevation
—extremely narrow
—low- or moderate-density single-family homes
—military (Coast Guard) property

A quick sampling of salient property damage risk attributes of islands in the Carolinas reveals that large differences may exist between adjacent islands and even on the same island. Hence the principles of hazard risk mapping and property damage mitigation must be applied in the context of recognition that no two barrier islands or coastal zones are alike.

Beach; the mainland coast of Myrtle Beach and the Grand Strand; and North Carolina's Sunset Beach) and North Carolina's wide range of barrier islands and associated communities (see tables 9.1 and 9.2). Two areas, in particular, were included to show the range of community mitigation responses, namely the northern Outer Banks (e.g., Kill Devil Hills and Nags Head, North Carolina), and the communities on Bogue Banks, North Carolina. The examples are dispersed throughout chapters 5, 6, and 7, but table 9.1 provides a short recap, and Bogue Banks is presented again for overview and comparison.

Table 9.2   Illustrations of Risk Assessment and Property Damage Mitigation Concepts in the Carolinas

Soft Stabilization
The Sand Commandments
Adding sand to the beach
    "Free" beach replenishment-BB, TI
    Beach replenishment-GS, CH
    Beach bulldozing-TI, NH, BB
Increasing sand dune volume
    Sand fencing and dune building-BB, SB
    Other examples of dune building-SB, PI
    Plugging dune gaps-C
Adding sand to interior of island
    Building interior dunes-BB

Building Relocation
Relocation philosophy-CH
Nags Head: Town on the move-NH
The 10/100-year relocation concept-GS
Long-term relocation planning-BB, NH

Hard Stabilization
Shore parallel-GS, FB, CH
Shore perpendicular-FB, PI, CH

Vegetation
Lagoonside marsh planting-BB
Maritime forest and Hurricane Hugo-PI
Unnecessary removal of forest-BB

Modification of Development and Infrastructure
Blocking beach access roads-PI
Re-orientation of roads-FB

Zoning, Land-Use Planning
Recognize hazard areas and avoid
    Tidal inlets
        Inlet hazard area-Historical inlet-FB, BB
        Inlet hazard area-Modern inlet-BB, TI
        Inlet hazard area-Modern inlet dynamics-TI
        Inlet hazard area-Newly formed inlet-PI
    "Swashbuckling"-GS
Setback for protection-BB
Deep property lots for future relocation-BB, NH
Unwise building location-major structure-FB
Living with an accreting beach-SB
Offshore rubble-GS
Construction quality and hurricane damage-SI, IOP
Dune gaps or breaches increase island interior flooding-NH

Location Abbreviations:
FB=Folly Beach, SC
GS=Grand Strand, SC
BB=Bogue Banks, NC
TI=Topsail Island, NC
PI=Pawleys Island, SC
SB=Sunset Beach, NC
SI=Sullivans Island, SC

IOP=Isle of Palms, SC
NH=Nags Head, NC
CH=Cape Hatteras Lighthouse, NC
C=Corolla, NC

Assessment of hazards and property damage mitigation recommendations for several coastal communities from a wide geographic spread is intended to broaden the scope and applicability of this concept. Figure 9.1 highlights the locations referred to in this chapter, and table 9.3 provides a quick overview of the sample communities.

Keep in mind that the risk mapping criteria are based on the property damage risk potential from a moderate category 3 hurricane. For almost any community, everything is at extreme risk in a larger hurricane, especially a category 5.

9.1 Index map showing the locations of the communities selected as examples of risk maps.

Southampton, Long Island, New York

The township of Southampton, New York, is located on the south shore of Long Island in Suffolk County, approximately 80 miles (130 kilometers) east of New York City. Southampton Township includes the villages of Westhampton Beach, Quogue, Hampton Bays, Southampton, and Bridgehampton and occupies the western half of the south fork of Long Island, covering a total of 25 miles (40 kilometers) of the Atlantic shoreline.

The northern half of the town lies on the Ronkonkoma terminal moraine, deposited during the most recent glacial episode. The moraine consists of till, a poorly sorted mixture of clay, sand, and gravel that is highly erodible. The area has relief of over 200 feet (60 meters) at its highest points. This glacial moraine is well forested and is an important groundwater recharge area. The area between the moraine and the Atlantic Ocean is a glacial outwash plain. The outwash consists of well-sorted sand derived from the Ice Age glacial melt waters and slopes gently seaward. The rich agricultural soils of the outwash plain led to early rural development. Peconic Bay lies north of the moraine.

The nature of the Atlantic shoreline changes significantly from east to west in the area (fig. 9.2). The eastern half of the township is a headland or mainland-type shoreline which extends past the eastern town boundary to Montauk Point at the eastern tip of Long Island.

Within the village of Southampton the shoreline changes to a barrier coastline that extends westward to New York City. Two stabilized inlets, Shinnecock and Moriches, divide the barrier islands in Southampton (see fig. 9.2). Because of the low tidal range and

Table 9.3   Characteristics of Case Study Sites in Chapter 9

	Wave Energy	Storm Surge	Type Storm	Coast Type	Engineered/ Structures
Southampton, NY	High	Moderate	Hurricanes/ NEasters	Low/narrow	Yes
Bogue Banks, NC	High	Moderate	Hurricanes/ NEasters	One island Wide/high Dense forests	No
Fernandina, E FL	Moderate	Moderate	Hurricanes/ NEasters	Wide/very high Dense forests	Yes
Venice Beach, SW FL	Moderate	High	Hurricanes/ SWesters	Low/narrow	Yes
Wakulla Co., W FL	Zero energy	Extreme	Hurricanes	Low Forests	No
Galveston, TX	Moderate to low	High	Hurricanes/ SWesters	Low/open/wide	Yes

high wave energy environment, the barriers exhibit a wave-dominated morphology. The islands are long and narrow with few tidal inlets, similar to northern North Carolina. Moriches Inlet marks the western boundary of Southampton.

Very narrow stands of salt marsh grow in the lagoons between the barrier islands and the mainland. The two lagoons in Southampton (Shinnecock and Moriches Bays) are extremely shallow and several miles wide. Salt marsh vegetation only grows in two general regions: on overwash fans on the back side of the islands and in small protected coves within the bays.

The net littoral drift in Southampton is from east to west. Most of the sediment in the system is derived from the eastern headlands, but there also is evidence of an offshore source (the continental shelf) because the volume of sand transport on the beach increases to the west.

Numerous ponds and small embayments have formed in the eastern part of the town where former outwash streams were closed off by the development of baymouth bars. In fact, before Moriches Inlet formed in 1931, the south shore extended unbroken by inlets from Montauk Point to Fire Island Inlet. However, Mecox Bay and Sagaponack Pond are frequently inundated by storm tides. As sea level continues to rise, permanent inlets will likely form at these locations.

Southampton has endured a long record of brutal hurricanes and northeasters. The most destructive storm of this century was the hurricane of 1938, which wiped out development along the entire barrier coast and opened Shinnecock Inlet. The inlet was subsequently maintained by dredging and later by jetties.

The village of Southampton was founded in 1640 and became the first English settlement on Long Island and the first European settlement on the eastern end (the Dutch had previously settled in what is now Brooklyn). The indigenous Shinnecock Indians showed the early settlers the value of the region's rich coastal resources—shellfish, fishing, and even whaling. Until this century the local population consisted largely of baymen who depended on the sea for their living.

Since World War II the population on Long Island has exploded as New York City's suburbs sprawled eastward. Southampton has resisted the suburban development that has enshrouded the rest of Long Island. Today it depends more on its reputation as a wealthy, exclusive resort area than on its coastal resources. The development pattern on the oceanfront consists primarily of large, expensive, secluded homes. Commercial development on the beach is largely absent.

Southampton Risk Assessment and Mitigation Recommendations

The barrier island between Shinnecock and Moriches Inlets is all classified as extreme risk (see fig. 9.2). The elevation of the island is mostly below 10 feet (3 meters), and the V zone covers the entire width of the island. Erosion is severe due to two prominent structures: the jetties at Shinnecock Inlet and the groin field in Westhampton Beach.

The Shinnecock jetties block most of the littoral transport to these beaches. Storm overwash occurs frequently within 2 miles (3 kilometers) of the jetties. Increased erosion rates in Westhampton Beach after the construction of the jetties prompted the construction of a groin field which terminates 3 miles (5 kilometers) east of Moriches Inlet. The 21 groins were completed in 1977. A devastating scenario has unfolded to the west of the groins as erosion accelerated even more. While the groin field has stabilized the beach and protected homes within its limits, almost 200 homes have washed away within a mile of the last groin and a new inlet formed! This portion of the island is now almost barren of houses and vegetation, a mere sandbar separating the Atlantic from Moriches Bay. Development is coming back, however, as the citizens of West

Hampton Dunes successfully sued the Corps of Engineers, forcing them to return the shoreline to its original position through beach replenishment (your federal tax dollars at work). The effects are encroaching further and further westward, and every storm season sees more homes disappear.

East of Shinnecock Inlet a 3-mile-long (5 kilometers) barrier spit extends from the headland area. Because sand is impounded on the east side of the jetties, the beach within a mile and a half (2.5 kilometers) of the inlet is stable but is still rated as extreme risk. Continuous dune ridges range from 15 to over 30 feet (4.5 to 9 meters) high, greatly reducing the risk of overwash, but a few gaps do exist and these should be avoided. Secondary dunes add width to the island, but elevations on the back side of the island are low. The remainder of the barrier spit is eroding and has no secondary dunes, so it too falls into the extreme risk zone to be avoided (see fig. 9.2).

The mainland portion of the town's shoreline is eroding at a considerable rate, which increases from west to east. Although the dune crest is over 20 feet (6 meters) in some areas the dunes are severely scarped. In many areas, however, the dunes are low and do not have enough width to withstand a high storm surge. The bars blocking Mecox Bay and Sagaponack Pond are particularly vulnerable. They overwash during most storm tides and will very likely become permanent inlets as sea level rises. Beachfront property is at extreme

risk and should be avoided (figs. 9.3 and 9.4).

The margins of inland bays and ponds are floodable and are high-risk zones. Areas shown as moderate risk are generally open fields that are agricultural lands, but some are occupied by recently built subdivisions. The remainder of the town is low risk. Because of the slope of the mainland, low-risk sites can be found close to the ocean, but as erosion continues, these areas will be subject to increased risk.

The area of the town north of Montauk Highway was not mapped but is virtually all low risk. Likewise, the mainland portion of the town that is separated from the barrier island is also low risk, except for low-lying land adjacent to the bay.

Bogue Banks, North Carolina

Bogue Banks is over 25 miles (40 kilometers) long and over 1 mile (over 1.5 kilometers) wide at its widest and is oriented essentially east-west. The island is bordered on the east by Beaufort Inlet and to the west by Bogue Inlet (see figs. 6.2 and 9.1). High elevations characterize the island in places, with dunes reaching over 35 feet (11 meters) high on the western end. Even the relatively narrow, lower central part of the island is higher in elevation than most other North Carolina islands (see fig. 3.7). These positive elevations are less vulnerable to hazards, as reflected in figure 9.5. The storm surge from a category 5 hurricane would flood only about half of the western end of the

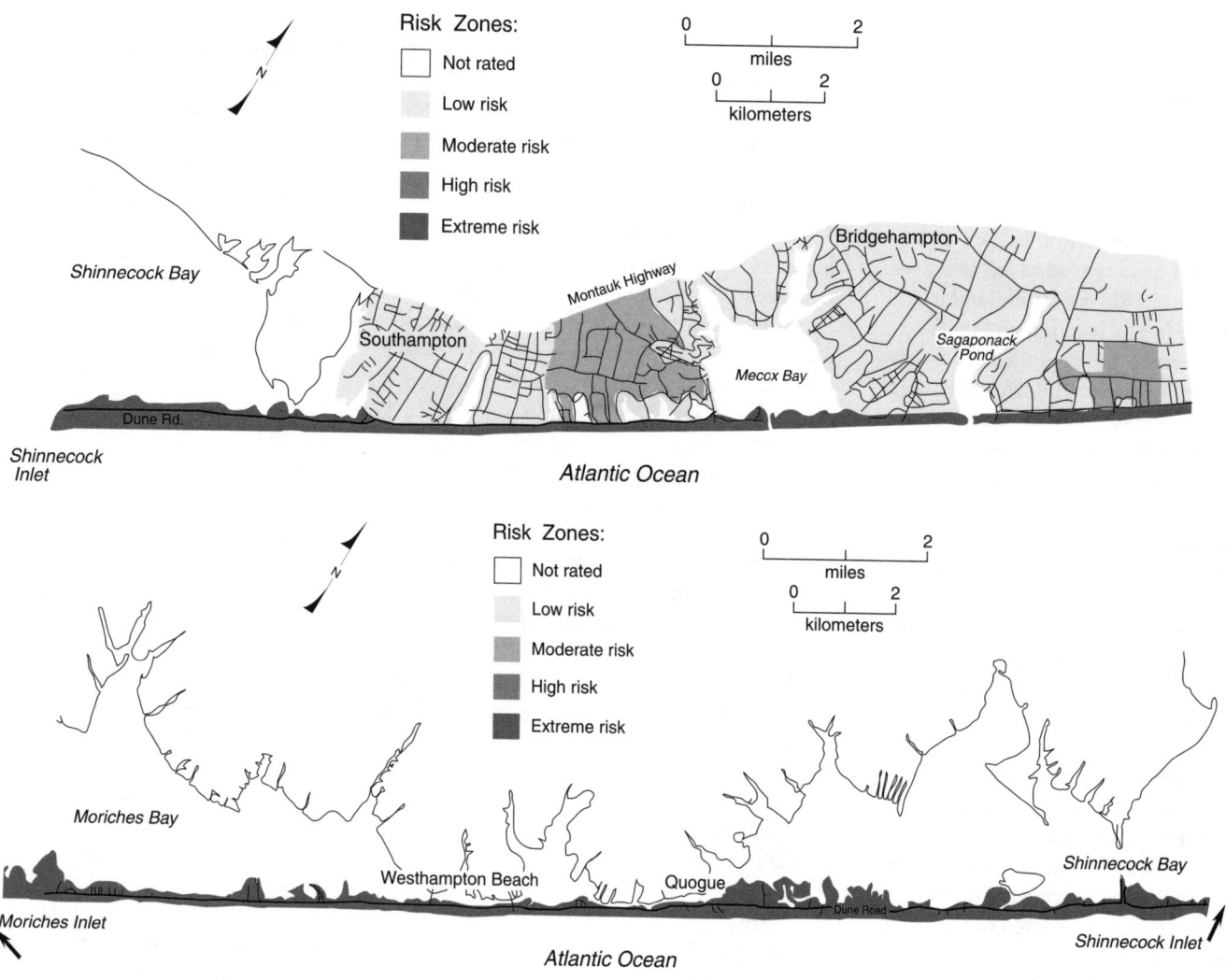

Risk Zones:
- Not rated
- Low risk
- Moderate risk
- High risk
- Extreme risk

Shinnecock Bay

Bridgehampton

Montauk Highway

Southampton

Mecox Bay

Sagaponack Pond

Dune Rd.

Shinnecock Inlet

Atlantic Ocean

Risk Zones:
- Not rated
- Low risk
- Moderate risk
- High risk
- Extreme risk

Moriches Bay

Shinnecock Bay

Westhampton Beach

Quogue

Dune Road

Moriches Inlet

Shinnecock Inlet

Atlantic Ocean

9.2 Risk map for Southampton on the south shore of Long Island, NY.

island, though all of the eastern half would be covered (NCDEM, 1987).

Some of the highest sand dunes on the East Coast and the largest in North or South Carolina south of Jockey's Ridge, North Carolina, can be found on Bogue Banks. The large dunes on Bogue Banks are testimony to the large volumes of sand moving ashore in this area, although the exact reason for such extensive sand movement in unknown. Part of the answer is certainly the east-west orientation of the island, making it perpendicular to the predominant wind directions. This orientation favors processes that bring a lot of sand onshore, compared to a very low-elevation island such as Core Banks, which is oriented north-south, parallel to dominant wind directions (Godfrey and Godfrey, 1976). The total answer is almost certainly more complex than that, as Jockey's Ridge, Nags Head, North Carolina, is an enormous sand dune and it is located on a north-south–trending shoreline.

Bogue Banks has five separate municipalities on the island (fig. 9.5), each with a different philosophy about "living with the shoreline." In addition, the island can be divided into three distinct parts geologically. The eastern third is low to moderate elevation and wide. The central third is low elevation and narrow. The western third is extremely high elevation (by barrier island standards) and very wide.

Bogue Banks Risk Assessment and Mitigation Recommendations

Many specific sites on Bogue Banks were discussed in chapters 5, 6, and 7. Bogue Banks is important to the study because it provides the best example of reduced risk attributable to natural island character (e.g., high elevation, large amount of sand in the system, and good forest cover) resulting in more moderate- to low-risk sites than other islands studied. Hurricane Hazel, a category 4 storm, visited Bogue Banks in 1954, the last hurricane of that magnitude to hit this portion of North Carolina. Hazel opened several new inlets along Bogue Banks, which, though they were artificially closed, can still be recognized today in the form of small tidal deltas in the sound and low-elevation spots along the main island road (see fig. 7.9). Hazel's impact on specific sites should have served as a warning. It did not!

Overall, Bogue Banks has one of the lowest risk ratings along the East Coast of the United States. Of course, no island is completely secure from the ravages of a major storm. A great deal of the natural protective capability of Bogue Banks has been destroyed by development, making even this lower risk island an invitation for disaster in some places. For example, Bogue has a significant sand volume, one of the keys to reducing property damage. A large sand volume means a wide and high island. Island elevation and width allow siting of development back and up from the water, two of the best ways to mitigate against storm

9.3 A homemade rubble seawall protects this Southampton house.
9.4 Many of the buildings in Southampton, NY, at immediate risk to erosion are large and costly.

damage. Unfortunately, in some places, much of the sand was removed, compromising the protective capacity. A study by Stanczuk (1975) documents some of the early effects of development on Bogue Banks's natural environment. The impact of development on the maritime forest resources on Bogue Banks is discussed by Lopazanski (1987) and Lopazanski, Evans, and Shaw (1988). Recommended mitigation repairs for such damage

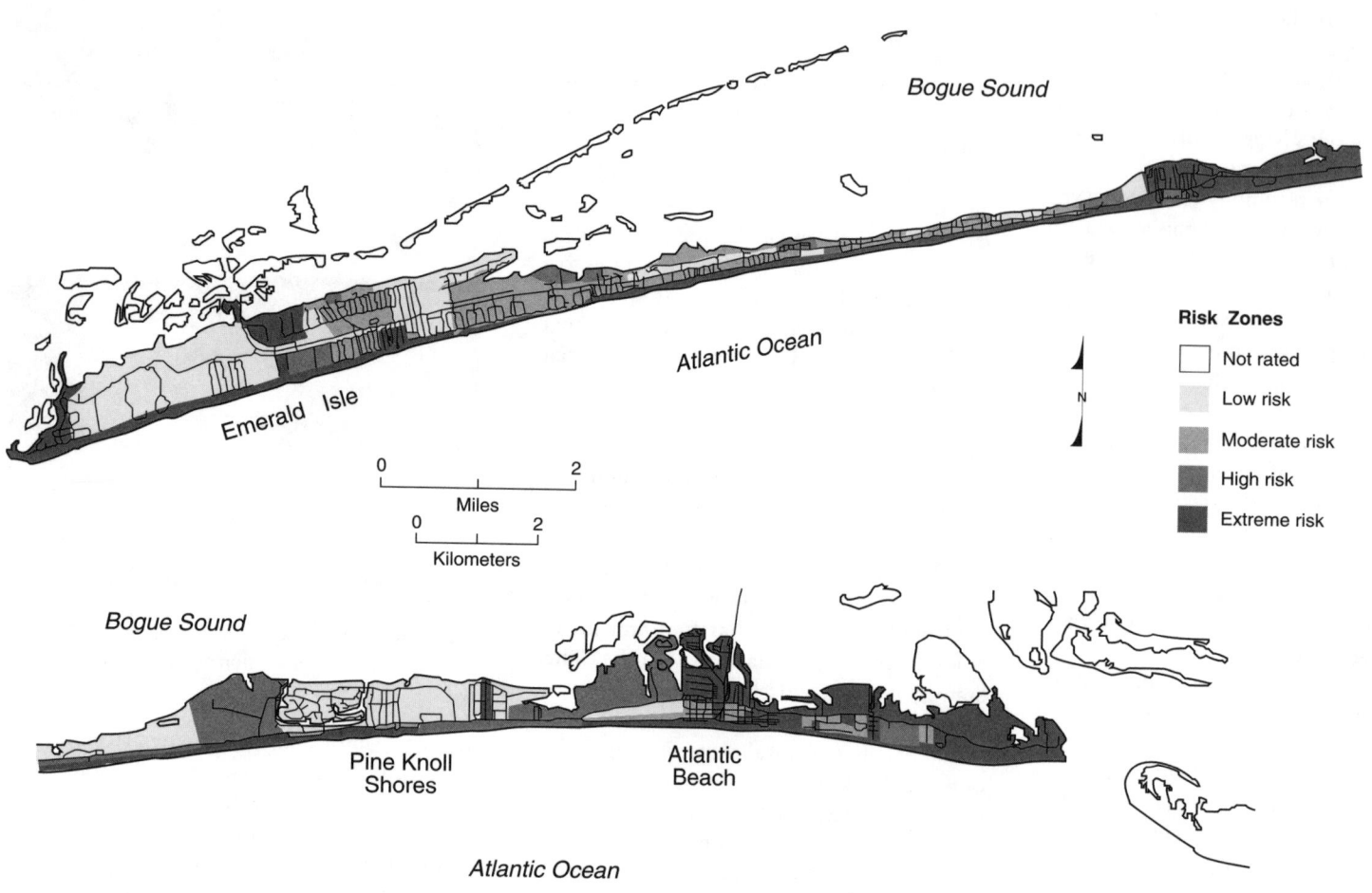

Risk Zones

☐ Not rated

Low risk

Moderate risk

High risk

Extreme risk

Bogue Sound

Atlantic Ocean

Emerald Isle

0   2
Miles

0   2
Kilometers

Bogue Sound

Pine Knoll
Shores

Atlantic
Beach

Atlantic Ocean

9.5 Risk map for Bogue Banks, NC. This island consists of several communities. Differences in city hall viewpoints make an overall beach and island management plan difficult to come up with.

are given in the generalized mitigation map for Bogue Banks (see fig. 6.2).

Numerous other examples of the destruction of the island's natural protective qualities exist on Bogue Banks. Very large dunes protected the community of Atlantic Beach, but they were totally removed from the heart of the city for ease of construction of buildings and for roads. Other high dunes were removed and dunes were notched for roads. Extensive protective maritime forest still exists on the island, but in many places the forest has been completely cleared for development. Locations of historical inlets are documented, yet development is continuing in these locations, which are likely to be sites of future inlets. And although modern building codes allow structures to withstand high winds and modest waves, many older buildings on Bogue Banks were built before building codes were established.

The list of trouble spots goes on. The response must be to repair the damage, encourage restoration of natural dynamics, and prevent further loss of dunes and forest (see fig. 6.2).

Atlantic Beach (see fig. 7.8) represents the portion of the island most modified by humans. The natural contours and environments have been greatly altered or obliterated. Marsh fill, finger canals, and septic tanks further detract from the island's natural character. The lagoon side of Atlantic Beach is mostly marsh fill. Several finger canals were dredged and the spoil dumped on the marsh to fill for home

sites. These locations are very low in elevation and certainly will be flooded even by a modest storm. Little can be done to improve these sites except for elevating the houses on stilts (pilings) above potential flood danger. Elevation will protect these buildings from rising water, but not from wind and wave damage or storm-surge ebb scouring of the stilts.

The far eastern end of Bogue Banks is occupied by Fort Macon State Park, which will not be developed much beyond its present status. The park's dunes, beach, and forest will remain in a relatively natural state. The actual fort area has been damaged by past storms. In fact, the earlier stone Fort Hampton, which protected the inlet during the War of 1812, was destroyed by a severe hurricane in the early 1800s. Private property between the park and the Atlantic Beach town limits has also suffered in past storms. This area of relatively low elevation is susceptible to overwash, a high rate of shoreline erosion, storm-surge flooding, and active sand dune migration, which minimizes the number of sites of even moderate risk for development. Lower risk sites are located near the center of the island where shrub stands and maritime forest indicate island stability. Any structures along this stretch that are above the 100-year storm-surge level are not likely to be flooded.

The eastern end of Bogue Banks, including the town of Atlantic Beach, has been the beneficiary of two beach replenishment projects at no cost to the community because the sand

was obtained from navigation channel dredging (see chapter 5; fig. 5.22).

Mitigation recommendations for Bogue Banks rely on the PAR approach: preserve, augment, and restore. The dense maritime forest, still extensive in places, should be preserved as much as possible. The same goes for the high dunes.

Fernandina Beach, Florida Atlantic Coast

The community of Fernandina Beach is located on Amelia Island, the northernmost island on the eastern coast of Florida and the southernmost of the "Sea Islands" (see fig. 9.1). Amelia Island is located about 11 miles (18 kilometers) northeast of Jacksonville and is bounded by the St. Mary's Inlet to the north and St. George Inlet to the south. The island is some 8 miles (13.5 kilometers) long, with an average width of roughly 1.8 miles (3 kilometers). Elevations at the back of the island can exceed 50 feet (16 meters), while the majority of the island's interior is well above 15 feet (4.5 meters), with an average elevation between 10 and 16 feet (3 to 5 meters). The highest dune on the island rises more than 54 feet (16.5 meters) above mean sea level. These unusually high elevations are found entirely on the northwestern portion of the island.

The present Amelia Island represents the merging of the modern barrier island with its more ancient predecessor (Parchure, 1982). The majority of the island is composed of

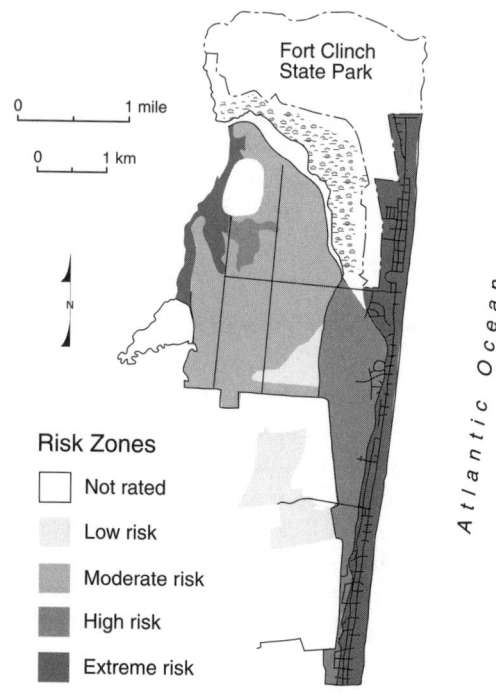

Risk Zones

☐ Not rated

☐ Low risk

☐ Moderate risk

☐ High risk

■ Extreme risk

9.6 Risk map for Fernandina Beach, FL.

reworked Pleistocene sediments that made up the ancient barrier island. Atop this Pleistocene core is a sliver of modern Holocene sediment (Parchure, 1982).

The oldest development in Fernandina Beach dates to 1567 and is located on the back side of the island along the Amelia River (Parchure, 1982). Elevations there typically exceed 40 feet (12 meters), an elevation gener-

ally safe from flooding. The city of Fernandina Beach has grown outward toward the beach extensively since the passage of Hurricane Dora in 1964.

The northern portion of Fernandina Beach (north of Atlantic Avenue) is composed mainly of single-family dwellings, most of which are not elevated on stilts. The base flood elevation in this area ranges from 10 to 16 feet (3 to 5 meters) along the beach itself. The first row of houses sits behind a low-elevation (5 feet, 1.5 meters), unvegetated dune ridge, often less than 10 feet (3 meters) wide. There are as many as 11 rows of houses and as few as 2 rows between the ocean and the densely forested dune ridges. The southern portion of Fernandina Beach (south of Atlantic Avenue) is composed of single-family dwellings, two- to seven-story condominiums, and small stores and businesses. Most of these buildings are not stilted and are located directly behind or on top of the first row of dunes. The majority of this development lies along Highway 105. For this reason, there are typically only two to four rows of buildings between the ocean and the densely vegetated dune ridges. The base flood elevation in this area also ranges from 10 to 16 feet (3 to 5 meters). The central region of the island hosts the Fernandina Beach municipal airport, which occupies approximately 0.6 square miles (1.5 square kilometers).

Amelia Island exhibits a highly engineered shoreline. The earliest coastal engineering projects began in 1881 with construction of

five spur groins around Fort Clinch (Parchure, 1982). In 1882, The U.S. Navy began work on an 11,000-foot-long (3,350 meters) jetty with a height of 6 feet (1.8 meters) above mean low water. The jetty stabilizes the St. Marys entrance. A year later, in 1883, two additional groins were added in the Fort Clinch area (Parchure, 1982). Once the jetties were completed, dredging programs were necessary to keep the inlet open. Sediment removal ranged from about 22,000 to 200,000 cubic yards (17,000 to 150,000 cubic meters) per year (Parchure, 1982). In 1953, the city of Fernandina constructed eight groins at intervals of about 425 feet (142 meters) extending north of Atlantic Avenue (Stevens, 1960). In 1964, funds were provided for 3.4 miles (5.4 kilometers) of revetment (Pilkey et al., 1984). In the years 1978, 1982, 1987, and 1988 sand dredged from federal navigation projects was used for replenishment (Hobbs, 1988), though no record of the total volume of material emplaced is available.

Fernandina Beach Risk Assessment and Mitigation Recommendations

The coastal risk zones for Fernandina Beach are shown in figure 9.6, which divides the community into several areas. Blank areas on the figure represent undevelopable lands which have therefore been left unranked (Fort Clinch State Park, the airport, industrial areas, and marsh). Low-risk areas include two small areas

9.7 The beach at Fernandina Beach, FL. Photo by Amy Reesman.

south of town, distinguishable by high-elevation dunes, dense forest, and shrub thicket. Moderate-risk areas include the older developed section of the town. High-risk areas include most of the state park area and a large section of the southern portion of town parallel to the coast. Extreme-risk areas are confined basically to the oceanfront. Three distinct portions of Fernandina Beach are rated as extreme hazard (fig. 9.6): the public beach and one zone each to the north and south of the public beach.

The extreme-risk area north of the public beach lies behind an unsubstantial man-made dune ridge ranging from less than 3 to 5 feet (1 to 1.5 meters) in height. The majority of these residences have not been constructed on stilts, although some structures utilize the first floor as garage space only. The artificial dunes in this stretch are small, unstable, and not continuous. In addition, low-lying roads are perpendicular to the shore and encourage overwash and storm-surge ebb. Several houses are located too close, less than 130 feet (40 meters), to the ocean. The sparse vegetation offers no protection (fig. 9.7).

Mitigation recommendations for this area are limited. There is little or no space to build dunes in front of houses, so relocation becomes a more reasonable alternative. If houses are not relocated, the volume of the artificial dune must be augmented, increasing its height and width so that it will offer more protection during storms. But who will pay for this? Dune

gaps must be plugged and the dunes vegetated for increased stability. In addition, a program should be initiated to plant and encourage native vegetation, especially shrubs and forest vegetation. Some streets (for example, Seventh and Ninth Streets) can be completely closed off, filled in, and revegetated. No private drives are located in these areas so access is of no concern. Other streets (for example, First, Second, Third, Fifth, and Eighth Streets) can be blocked with sand at their ocean terminus.

Fernandina public beach is also an extreme-risk area (see fig. 9.6). Here the road perpendicular to the shore will allow overwash penetration and storm-surge ebb. The rock revetment will cause beach narrowing over time. Sparse vegetation offers no protection. The corresponding mitigation recommendations for this area are to block the road gap, begin planning for replenishment in the future to maintain the recreational beach, and develop a planting program.

South of the public beach is another extreme-hazard area. Development in this area is distinguishable from development north of the public beach in that it lies on top of or directly behind a lightly vegetated natural dune ridge of moderate size. The development consists mainly of one- and two-story, nonstilted, single-family dwellings. The only beachfront condominiums in Fernandina Beach are located along this strip of Highway A-1-A. Dunes in this area are sparsely vegetated and are therefore less stable. Where houses are

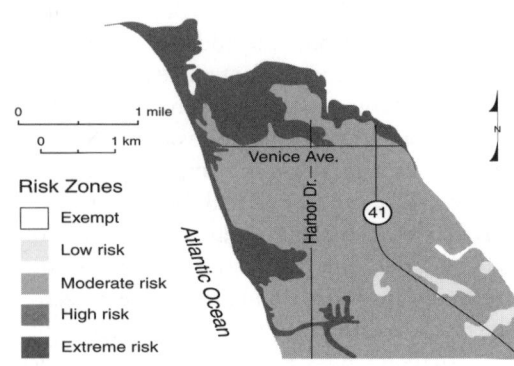

9.8 Risk map for Venice Beach, FL.

built on top of the dune or directly behind it, the dune offers no protection, except for some elevation. Driveways and roads perpendicular to shore encourage overwash and storm-surge ebb. Many houses are located less than 200 feet (60 meters) from the ocean. On-beach parking encourages erosion and creates a dune gap.

Mitigation recommendations for this area include planting and encouraging native vegetation, especially along the dune ridges, for stabilization, removing beach parking access, and filling the dune gap.

Venice Beach, Florida Gulf Coast

Venice Beach is located approximately 60 miles (100 kilometers) south of Tampa Bay, facing the Gulf of Mexico (see fig. 9.1). Venice is bounded by Venice Inlet (jettied) to the north and Stump Pass to the south. Elevations reach as high as 40 to 50 feet (12 to 15 meters) above mean sea level, with a majority of the land between 10 and 20 feet (3 to 6 meters) above mean sea level.

The entire Venice area is heavily developed and few areas with dense vegetation remain. The first row of buildings nearest the shore varies as follows: buildings from the jetty (Tarpon Center Drive) to just south of Whitecap Circle are one- to three-story apartments and condominiums built at grade (without stilts); south of this area, to the public beach (the Esplanade), is a variety of high-rise hotels and apartments, most ranging from four to eleven stories; and just south of the public beach is a variety of single-family dwellings, bounded at the south by an eleven-story high-rise. The second row of buildings consists primarily of one- to two-story apartments and condominiums, with the exception of a three-block area with low-rise (three- to six-story) apartment build-

9.9 The vertical bluff indicates that this house in Venice Beach, FL, is in serious erosion trouble. Photo by Amy Reesman.
9.10 A rock revetment on Venice Beach, FL. The swimmers exiting and entering the water clearly have a hazardous task. One of the major negative effects of rock revetments is the swimming hazard both from the standpoint of access to the water and the problem of stray, underwater rocks thrown offshore by storm waves. Photo by Amy Reesman.

ings. Most of the other development in Venice consists of one- to two-story single-family dwellings or businesses, all of which are built at grade, not elevated on stilts or pilings. Several trailer parks are located about one mile inland. The Venice municipal airport, just south of Venice, occupies approximately one square mile (2.6 square kilometers).

Shoreline engineering began in Venice in 1937 when two 650-foot-long (200-meter) jetties were constructed to stabilize Venice Inlet. In 1963, 19,000 cubic yards (14,500 cubic meters) of sediment dredged from the inlet was used in a replenishment program (Dixon and Pilkey, 1991). Three finger canals were cut into the bayside of the northernmost tip of Venice and seawalls and rock revetments were constructed almost the entire length of the beach (dates unknown). Beaches were again replenished in the years 1971–75 as well as in 1979 and 1980 and into the 1990s (Dixon and Pilkey, 1991).

Venice Beach Risk Assessment and Mitigation Recommendations

The coastal risk zones for Venice Beach are shown in figure 9.8. Low-risk areas include a few small high-elevation, forested hummocks well back from the shoreline. Moderate-risk areas dominate the community and constitute the bulk of the developed areas. Although there are no areas in Venice Beach rated as high risk, extreme-risk areas border the ocean-

front and the area around Roberts Bay in the northern part of the community.

Hard stabilization structures (jetty, seawalls, and revetments) along the Venice Beach shoreline will lead to beach narrowing or create a need for replenishment with time (figs. 9.9 and 9.10). Dunes are low or absent, and vegetation is generally sparse or nonexistent. Finger canals and the marina allow further inland incursion of storm waters. Many buildings are located less than 200 feet (60 meters) from

9.11 A beach scene in Venice Beach, FL, showing the shoreline cut back from the seawall protecting the high-rise building. Clearly this is a community with an erosion problem, and efforts to halt the erosion are having a negative effect on beach quality. Photo by Amy Reesman.

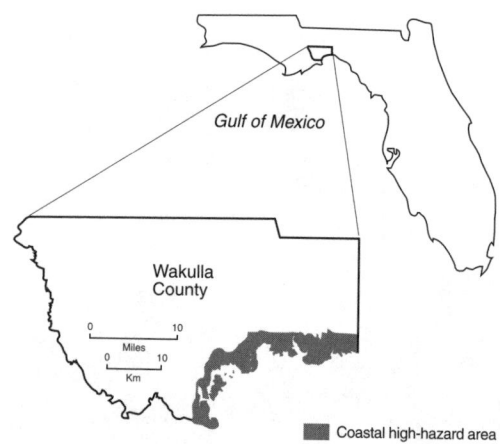

9.12 Location of Wakulla County, FL, showing the coastal high-hazard area in black. (Adapted from Wakulla County Comprehensive Plan, 1994.)

ocean, and roads perpendicular to shore will allow overwash and storm-surge ebb funneling. Large commercial high-rise structures may be more resistant to damage during storms, but damage to the buildings' contents and to windward and lower floors will be extensive (fig. 9.11).

Mitigation recommendations for Venice Beach are to add sand volume to dunes (frontal and interior), vegetate with native plants, and encourage shrub and forest growth. Relocation will be a necessity at some point. The community should be evaluating the possibility of demolishing and rebuilding if the economics of moving intact aren't feasible. Block some

shore-perpendicular roads with sand to inhibit overwash and storm-surge ebb.

The Venice public beach area is a valuable natural resource. However, low dunes and sparse vegetation offer little protection. No mitigation recommendations are made for this basically undeveloped shoreline stretch, although future replenishment to maintain the recreational beach should be considered.

Wakulla County, Florida "Zero-Energy" Coast

Wakulla County is located about 21 miles (35 kilometers) south of Tallahassee on Florida's panhandle (see fig. 9.1) and is home to Wakulla Springs Lodge, a well-known resort. Wakulla County has no barrier beaches. This mainland coast (fig. 9.12) occupies a setting on the low-energy reentrant section of Florida's Gulf of Mexico coast. The region is distinguished by a very broad and gently sloping continental shelf, the overall concave shape of the shoreline, and the local embayment of Apalachee Bay. The very gentle slope contributes to the low wave energy conditions due to frictional damping of the waves as they travel across the shelf. The gentle slope and the embayed shape of the coastline allow for maximum potential storm-surge heights. Storm surges here theoretically can exceed 30 feet (9 meters). In addition, extremely low energy shoreline settings prevent formation of barrier islands. Little sandy beach property exists in Wakulla County, as most of the coastline con-

9.13 In the coastal area of Wakulla County, FL, the 100-year flood elevation is in excess of 30 feet (over 9 meters), which explains why this building (top photo) is on such high stilts. The explanation for the unusually high elevation of the 100-year flood level lies in the great width and shallow slope of the adjacent continental shelf. A very flat continental shelf forces storm-surge waters to pile up to high elevations. Homes not elevated to this height were built before federal guidelines were in place, accounting for houses of three different elevations (bottom photo).

sists of coastal marshes grading into swamps. The large influx of fresh water from Wakulla Springs and other sinking streams maintains a freshwater marsh ecosystem along portions of the shoreline in addition to the expected salt-water marsh ecosystem.

Wakulla County is sparsely developed; the entire county's 1990 population was approximately 15,000 residents. Due to the large quantity of land held by St. Mark's Wildlife Refuge, development along the coast is not extensive. The main coastal communities of Wakulla County are Shell Point, Oyster Bay, and Live Oak Island. In general, houses in all areas are found at three physical levels: (1) nonelevated, built prior to the FEMA-regulated coastal development; (2) elevated 10 to 12 feet (3 to 3.6 meters), built to original FEMA flood standards; and (3) elevated 30 feet (9 meters), meeting present FEMA flood regulations. It is not uncommon to see houses at all three elevations right next to each other (fig. 9.13). Obviously the lower elevation houses are at higher risk for property damage and, if damaged or destroyed, debris could be carried by wind or water and damage nearby properly elevated buildings.

The community of Shell Point is comprised almost entirely of single-family dwellings. The first row of houses is mostly stilted. Approximately half of the buildings in the community, however, are trailer homes, located along canals in the central portion of the developed area. The community of Oyster Bay is a small community made up entirely of single-family housing units, mostly elevated on stilts as is the first shore row of houses. Several mobile home units are interspersed among the permanent structures. Live Oak Island is also a residential area. The houses on the first row are stilted, but Live Oak Island differs from the other communities in that there are few mobile homes.

Shell Point displays the highest degree of engineering with its numerous finger canals both parallel and perpendicular to the coastline. Oyster Bay has a series of finger canals parallel to the coastline. The roads here are also parallel to the coast. Live Oak Island is the only community of the three that boasts a seawall. The seawall is approximately 3 feet (about 1 meter) high. The landward side of the island has individual docks which are kept operable by the use of a revetment. Live Oak Island also maintains a series of canals.

The reconnaissance evaluation shows that the low elevation, numerous finger canals, and high potential for storm surges combine to place the entire area at extreme risk from coastal hazards.

Wakulla County Risk Assessment and Mitigation Recommendations

Wakulla County's extreme-risk coastal zones are divided into three main sections—Oyster Bay, Shell Point, and Live Oak Island—based on similarities in engineering, development, and geology. General mitigation recommendations are made for the entire shoreline rather than for each individual zone.

In the community of Oyster Bay all front-row houses are stilted, as are many of the houses away from the shore. Mobile homes also are present in this community. Coastal hazards of Oyster Bay include finger canals that will allow further inland penetration of storm waters, sparse vegetation, lack of dunes, mobile homes that are easily damaged during storms, and houses built less than 60 meters from the ocean. In the community of Shell Point, houses are one to two stories and a majority of those on the first row are stilted. Many mobile homes are present. Coastal hazards of Shell Point include finger canals, sparse vegetation, mobile homes, no dunes, roads perpendicular to the shore, and houses built less than 60 meters from the water. On Live Oak Island most houses are stilted and there are no mobile homes. Coastal hazards include finger canals, no dunes, houses built less than 60 meters from the water, and rock revetment and seawall that will lead to long-term beach degradation.

General recommendations for the Wakulla County shoreline are simply to elevate all structures and initiate a relocation plan for frontal units. Adding sand volume (sand trucked in from a more landward source) to build dunes (frontal and interior) would help reduce the impacts of small storms.

Galveston Island, Texas

Galveston is one of the oldest cities located on a Gulf of Mexico barrier island. Early Galveston grew as a prominent port city and was a profitable location for merchants and entrepreneurs. By 1900 the population was about 40,000. But in September of that year disaster struck: The deadliest hurricane ever to strike the United States took more than 6,000 lives, the worst natural disaster ever, in terms of death toll, to strike the United States.

After the Great Storm, the city of Galveston undertook one of the largest engineering projects ever on the U.S. shoreline. By 1902 a 4-mile-long (6.4 kilometers), 17-foot-high (5 meters) seawall was in place to protect the city from future storms (see fig. 5.3). In addition, 12 million cubic meters (16 million cubic yards) of sand were pumped into the city to increase the elevation of the island (see fig. 6.3). The project lasted until 1906 and required tremendous sacrifices of the city's residents. Since the original wall was constructed, an additional 6 miles (10 kilometers) of concrete wall has been added, along with lines of revetment at its base to protect the wall.

Despite these costly efforts, Galveston remains extremely vulnerable to hurricanes. Although the wall has been successful against most of the subsequent storms of this century, a storm with the same magnitude as the 1900 hurricane would still demolish much of the city.

Two glaring hazards plague Galveston Island. First, the highest elevation on the island is less than 15 feet (4.5 meters). The only high ground available is behind the seawall in the old city where the grade was artificially built. Second, dune heights on the entire island are less than 10 feet (3 meters), and secondary dune development is virtually absent. These elements combine to create a great flood hazard from any significant storm surge (fig. 9.14).

Galveston Island is almost 30 miles (about 50 kilometers) long and is entirely incorporated as the city of Galveston, although the city proper covers only the eastern third of the island. The island's 2-mile (3-kilometer) width, unusually wide for a barrier island in a microtidal environment, is due to the fact that in the past few thousand years the island has had a large sand supply and has widened in a seaward direction.

Galveston Island Risk Assessment and Mitigation Recommendations

The eastern end of the island is mostly moderate to high risk (see fig. 9.14). Due to the seawall and the artificial grade built early this century, the land immediately behind the seawall is where the highest elevations are found, and so it is classed as moderate risk. The lower elevations are found toward the back side of the island, where the risk is high.

To the west, the island is flat and has a rural flavor, complete with cattle ranches. Many low-lying interior areas are covered by freshwater marsh. The dunes are low and narrow and vegetation consists of small shrubs and grasses typical of the semiarid climate; no developed forest has ever existed on the island. This half of the island is all rated as extreme risk or high risk (see fig. 9.14). The delineation is made at the boundary of the V zone and A zone. The area is easily flooded, easily overwashed, and has no dense vegetative protection. Hurricane Alicia passed over Galveston in 1983, destroying many homes in her wake (figs. 9.15 and 9.16).

A potential problem with the seawall and fill is that both water washed over the wall and rainwater are forced to drain to the back of the island because of the graded fill. This back-side flow adds to the flooding potential. And, as with most seawalls, Galveston has no beach in front of the wall. The western terminus of the wall is a fine place to observe the loss of beach. The beach is now displaced about 100 feet (30 meters) landward of the seawall, where it used to be several hundred feet seaward.

The positive result of the seawall has been property protection and halting shoreline retreat. In this respect, the wall is one of the most successful seawalls in the United States. The city probably would never have regained its footing without its protection. Still, major hurricanes such as Beulah, Carla, Celia, and

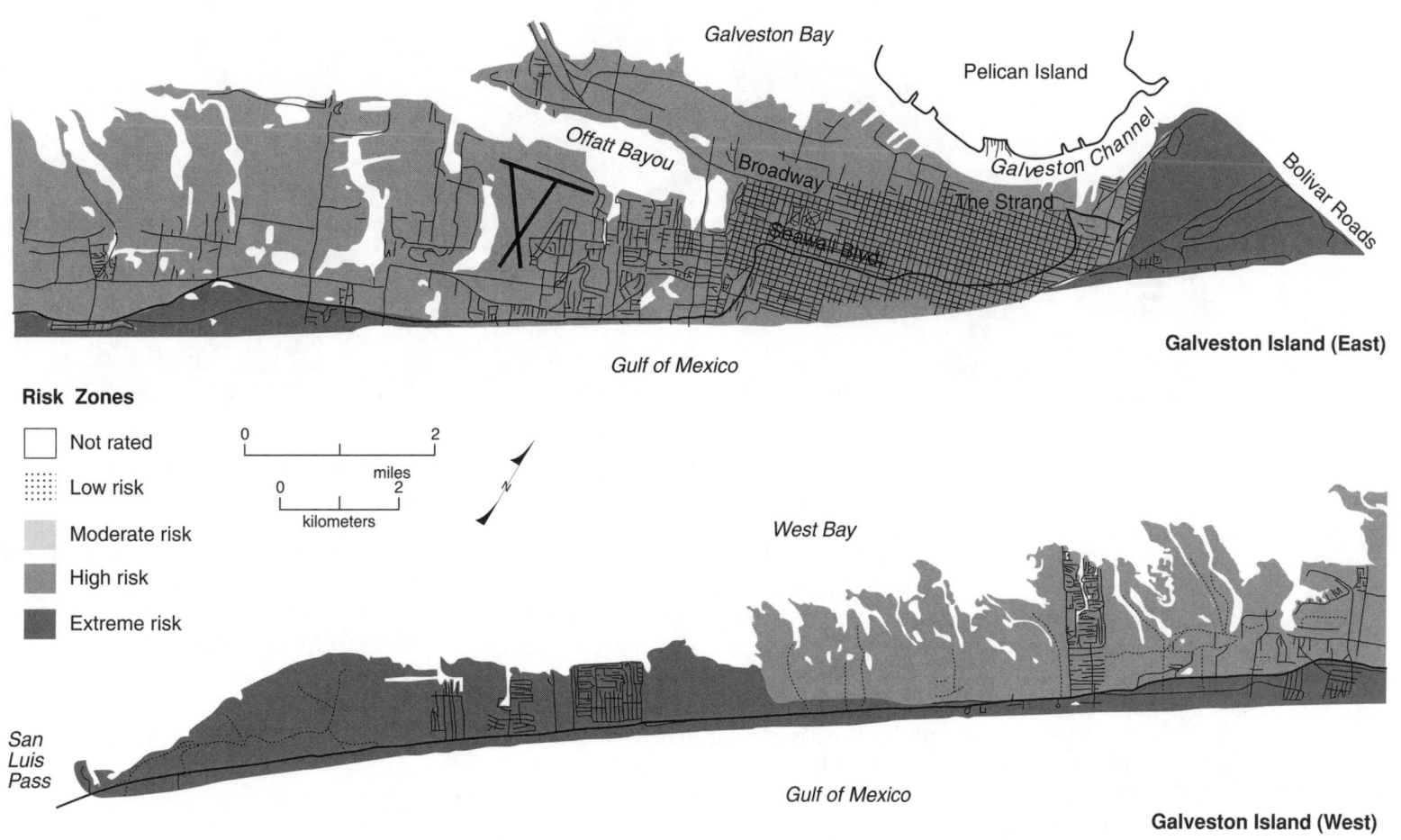

Galveston Bay

Pelican Island

Offatt Bayou

Broadway

Galveston Channel

The Strand

Bolivar Roads

Seawall Blvd.

Gulf of Mexico

Galveston Island (East)

Risk Zones

- ☐ Not rated
- ⠿ Low risk
- ▨ Moderate risk
- ▨ High risk
- ▨ Extreme risk

0   2
miles
0   2
kilometers

N

West Bay

San
Luis
Pass

Gulf of Mexico

Galveston Island (West)

9.14 Risk map of Galveston Island, TX.

9.15 Galveston hurricane damage from Hurricane Alicia (1983). Photo by Ernie Estes.
9.16 Galveston hurricane damage from Hurricane Alicia (1983). Photo by Ernie Estes.

Alicia have done significant damage. Additional protection is afforded to property because development is set back behind Seawall Boulevard, a four-lane highway. Fortunately, the loss of beach in front of the wall has been partially offset by sand impoundment behind the jetty at the east end of the island. This area of sand accumulation, called a fillet, provides a wide, flat, public recreation beach.

In 1995 the city's $4 million beach replenishment project rebuilt an artificial beach along the 4-mile (6.4-kilometer) stretch be-

tween Tenth and Sixty-First Streets. This project was the first public renourishment attempted in Galveston and took more than two years to be approved. The borrow site is less than a mile (1.6 kilometers) offshore and could have a large impact on sediment transport along the nourished beach or adjacent beaches. In June of 1995, Hurricane Allison churned by Galveston and, according to television reports, the beach replenished up to that time largely disappeared.

Commonality of Coastal Communities

These few examples, ranging from Long Island, New York, to Texas, ilustrate that each community must evaluate the hazards of their

coastal zone and tailor their property damage mitigation plans to meet specific community needs and property protection goals. The rules of the sea, however, are common for all coastal communities. Whether you are a public official, planner, or private property owner, you should apply the general principles of property damage mitigation.

At the beginning of the 1995 hurricane season, an Associated Press news article noted that two major hurricanes striking populated areas in the United States (e.g., Atlantic and/or Gulf coasts; California) in a single year could cause $50 to $80 billion in insured losses. Adding in a moderate earthquake could bankrupt the American insurance industry. A series of big storms in successive years could have the same effect. Such multiple events have occurred over the short term in spite of probability statistics to the contrary (e.g., multiple hurricanes in the Carolinas during the 1950s; Hurricanes Camille and Frederic in an 11-year interval; the floods along the Missouri and Mississippi Rivers in 1993 and 1995).

The probabilities of natural hazards and economics do not mix. As noted in chapter 1, the growth in population and capital investment in the coastal zone has passed a critical point in terms of risks from natural hazards and probable future property damage and loss. Hurricanes Hugo and Andrew clearly demonstrated the magnitude of the problem and illustrate the necessity for all communities to establish programs to reduce future property losses.

The several representations of our fictional Pandora's Island are used as a recurring theme throughout the book. It would be well for property owners and planners to review them (figs. 10.1 and 10.2). The island is seen first in its natural state and then as development comes (see chapter 2, figs. 2.1 and 2.4). Next is a drawing of the zones likely to be affected by various storm processes on Pandora's Island and Pandora's Island after the big one (see chapter 3, figs. 3.1 and 3.2). Using our risk mapping technique, the risk zones for Pandora's Island are drawn based on elevation, vegetation, and secondary factors such as landforms, inlet position, cover, sites of new inlet potential, potential ebb focus, and so on (see chapter 4, fig. 4.1). Placing development in hazard zones leads to the need for mitigation, but typically mitigation has meant doing something structural to the front of the island (walls, groins, beach replenishment) (see chapter 5, fig. 5.1). Our mitigation recommendations consider the larger coastal zone system, and we thus present a whole-island hazard mitigation plan for Pandora's Island (see chapter 6, fig. 6.1). Finally, Pandora's Fantasy Island represents a lower risk ideal development, preserv-

10.1 Pandora's Island. (a) The natural state. (b) Risk zones based on the island's characteristics before development. Risk refers to property damage potential in a moderate, category 3 hurricane hitting the island head-on. (c) Pandora's Island after being struck by a big storm.

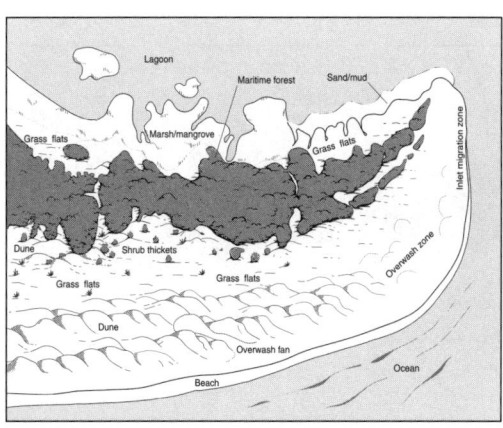

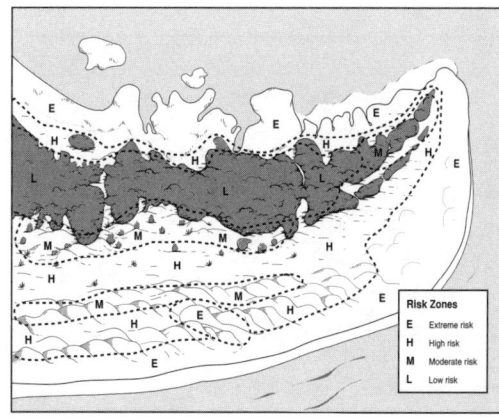

Risk Zones
E   Extreme risk
H   High risk
M   Moderate risk
L   Low risk

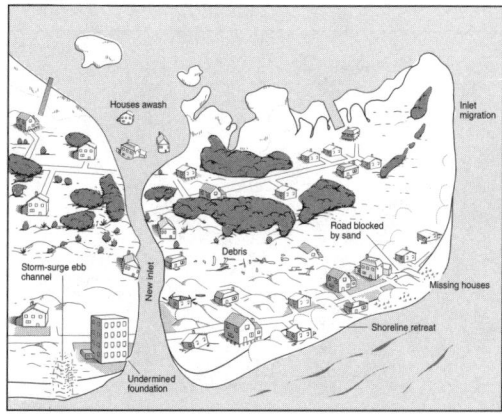

Lessons for Coastal Management

Post-storm observations of the impact of Hurricanes Gilbert (1988), Hugo (1989), Bob (1991), Andrew (1992), and Emily (1993) and several winter storms on developed shorelines helped define several principles or lessons learned regarding property damage mitigation (Thieler and Bush, 1991; Bush and Pilkey, 1994). These generalized conclusions are the basis of the rules of the sea presented earlier and related concepts that are restated here.

(1) *Wide beaches protect property.* The more beach available to absorb and dissipate storm-wave energy, the greater the possibility for mitigating damage to structures. The greater the distance between the zone of wave action and fixed construction the better. When beaches narrow, replenishment or soft stabilization is a means of widening them and increasing their "storm buffer" capacity. Beach replenishment is, however, expensive and temporary. As erosion threatens, a better alternative may be to move buildings.

(2) *Dunes protect property.* Sand dunes are often referred to as the "barrier" in barrier island, or "nature's shock absorber." The mass of dune sand may absorb and dissipate storm-wave energy, thus protecting buildings located behind dunes. Where dunes, rather than buildings, are available to absorb the impact of waves and storm surge, post-storm beaches are markedly wider. Dunes are the sediment reservoir, banked for a stormy day, that provide sand to the beach profile as it readjusts to storm-wave energy. Dune systems reduce overwash potential. In addition, interior dunes provide elevation for homesites, reducing flood damage potential. Building placement can take advantage of larger dunes to afford wind protection. Remember that dune width (and *dunefield* width) is as important as dune height.

Ideally, no dune on any portion of a barrier island should ever be removed. In most states, frontal dunes are protected, but interior dunes are not. Removal of interior dunes may lead to property loss and damage that could have been prevented if these dunes had remained intact. Killing or removing dune vegetation leads to reactivation of dunes by wind, creating blowouts and blowing sand that will be a nuisance, if not a hazard. Conserve sand resources. Add sand to island interiors and vegetate to stabilize and trap moving sand. Don't lose a grain of sand!

10.2 Pandora's Island. (a) Property damage mitigation on the ocean side. (b) Property damage mitigation on the island's interior. (c) Pandora's Fantasy, or the way it should have been done.

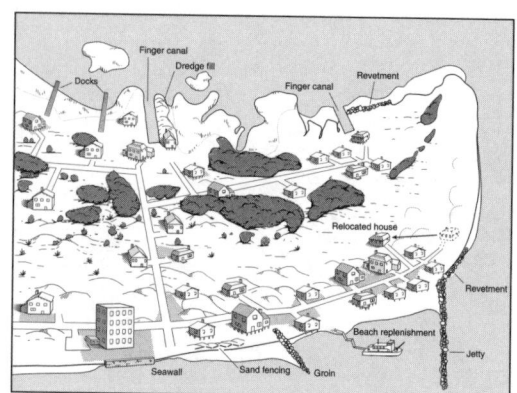

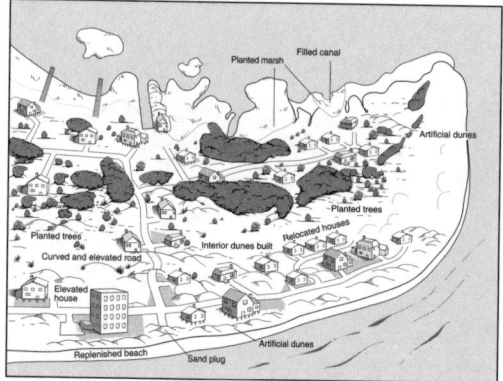

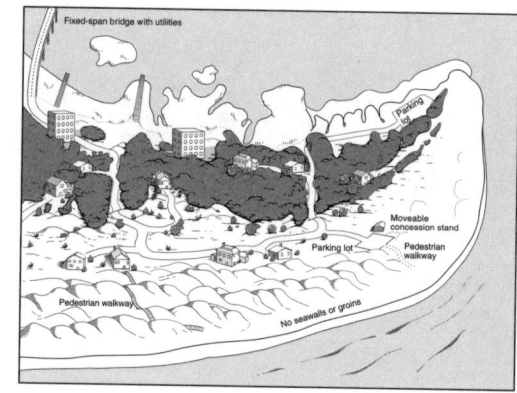

(3) *Vegetation protects property.* Overwash penetration and storm damage is noticeably greater where maritime forest is removed for development. This protective effect was well illustrated on Pawleys Island, South Carolina, where neighboring houses suffered vastly different degrees of damage from Hurricane Hugo depending on the degree of vegetation cover. Many houses located within the maritime forest were essentially untouched, except for some cosmetic damage; many houses built in cleared areas were destroyed. Similar effects were noted on the coast of the Yucatán Peninsula of Mexico after Hurricane Gilbert in 1988. As much forest as possible should always be retained (a practice exemplified by Pine Knoll Shores on Bogue Banks, North Carolina), and, where appropriate, reforestation of areas where trees were removed should be carried out. Once newly built dunes are stabilized with grassy vegetation, shrub and forest growth should be encouraged. Planting marsh grass on the backside of an island is preferable to bulkheading. Stabilizing cover of dune grass, marsh grass, maritime thicket, and mangrove is important to each respective environment.

(4) *Shore-perpendicular roads act as overwash and storm-surge ebb conduits.* Elevating and curving roads so they approach the beach at an oblique angle will reduce the extent and amount of overwash. Obliquely angled and elevated roads also provide a more difficult return path for storm-surge ebb flow, reducing scour potential. Simply putting a hump in the road at its beach terminus may help dramatically. A more effective method is a road-blocking scheme such as outlined for Folly Beach, South Carolina. Blocking these conduits in the interior of the island may be as important as plugging dune gaps. Roadbed material should never be obtained from the beach or the dunes.

(5) *Notches in dunes create overwash passes.* Notches cut in dunes for beach access, views, or construction sites are naturally exploited by waves and storm surge and by storm-surge ebb flows. Such notching can be avoided by constructing walkovers, elevating structures, and taking particular care during construction to avoid creating artificial passes. Where present, notches should be plugged or equipped with storm barriers that can be used to close these conduits prior to storms.

(6) *Overwash and storm-surge ebb is intensified when funneled by structures.* As storm-surge waters overwash an island and then return to the sea, driven either by gravity alone or in combination with onshore/offshore winds, existing structures may constrict the flow and reduce the cross-sectional area through which the water must pass. This tends to increase flow velocity, resulting in scour. This effect was well demonstrated by the storm-surge ebb channels observed on the Grand Strand shoreline after Hurricane Hugo and along the Yucatán coast after Hurricane Gilbert. Impermeable roads, drives, parking lots, and similar hard surfaces in constricted spaces between buildings prevent infiltration and add to the storm-surge ebb current effect.

(7) *Seawalls can protect buildings, but they also can cause narrowing of the beach, reducing both recreation and storm protection value.* Large seawalls are effective in protecting shorefront buildings from wave attack. But seawalls, as a rule, cause degradation and even eventual loss of beaches. Loss or narrowing of beaches reduces their storm buffering role and also may damage the tourist economy. Low seawalls on the Yucatán Peninsula and in South Carolina were, without exception, overtopped by the storm surge associated with recent hurricanes. Low seawalls in particular may serve to inhibit gradual shoreline retreat, but they afford minimal storm protection for property. Given such minimal tradeoff, seawalls are a poor choice for even short-term protection of property from storm surge, associated waves, storm-surge ebb scour, and flooding. Some coastal states now restrict or prohibit seawall construction on open ocean shorelines in order to prevent the loss of recreational beaches.

(8) *Setbacks protect.* Choosing a beachfront building site well back from the sea is the easi-

a

b

c

est and least costly method of property damage mitigation (fig. 10.3). As erosion threatens, structures can be moved. Most coastal states have some form of building setback regulations designed to reduce storm damage potential. Setbacks work over the short term because the beach storm buffer remains wide and wave energy from major storms is typically reduced by friction on land between the beach and the normal shoreline. If a shoreline is retreating due to long-term erosion, setback requirements are a short- to intermediate-term solution to property damage mitigation. However, federal and state programs provide incentives for relocation, either moving back on property or demolishing and rebuilding in another location.

(9) *Elevation protects*. Elevation, whether achieved by natural land elevation, infilling of a construction site, or by building on pilings, may be the single most important site-specific

factor in property damage mitigation (see fig. 10.3). Combined elevations of published FEMA 100-year-flood elevations plus predicted storm wave heights are incorporated into building codes for habitable first-floor elevations on virtually all U.S. barrier islands. Building codes, however, represent minimum requirements, and property owners may want to elevate even higher. But be aware that the advantage of elevation increases may be offset by the hazard of increased wind velocities in the "big one."

(10) *Proper community governance offers a degree of self-protection*. Development where building codes are enforced, and barrier island environments and processes allowed to operate, is less susceptible to property damage than development where failed structures become agents of additional destruction. Something as simple as keeping the subfloor area of stilt houses free of obstructions allows overwash to

10.3 These three photos from Hollywood Beach, FL, compare the property damage mitigation capabilities of seawalls, setback, and building above grade as illustrated during Hurricane Opal in October 1995. The three structures stand directly next to one another. (a) This structure is built on grade and is farthest seaward of the three. The seawall did protect the building from being undermined, but did little to prevent both floors of the structure from being completely destroyed. (b) Built above grade and set back slightly landward of the structure in (a), this building suffered minor to moderate damage to the first floor. The water and waves that completely destroyed the first floor in (a) washed mostly under this building. (c) Directly next to the building in (b), this three-story structure (left side of the photo) is set back significantly further landward and is at a slightly higher elevation. It suffered almost no structural damage due to waves. Photos by Craig Webb.

take place without damage to the building. In contrast, ground-level enclosures are smashed by storm-surge floods and waves directly damaging that building and possibly others with the debris from the damaged/destroyed enclosure. Know your building codes and see that they are followed for new construction. A home inspection of a previously built home will give insight into the quality of construction and adherence to codes.

(11) *Property damage mitigation must be applied islandwide.* The foregoing principles are inadequate if applied singly or only in one island subenvironment or one development. Traditional mitigation techniques that focused on the shoreline have not reduced damage to island interiors or, in some cases, shoreline property. A community that nourishes its beach but levels interior dunes has yet to grasp that a barrier island is an integrated system. Although recommended actions may be site specific, the total set of mitigation actions must be compatible and applied over the entire island.

The Coastal Processes Approach

Although the traditional approaches to property damage mitigation (legal regulation, construction requirements, and protective engineering) must continue to be applied, the main purpose of this book is to promote recognition of mitigation approaches based on the physical processes active within coastal environments.

The theme is to develop a *coastal processes approach* to property damage mitigation, the basis for recognition and mapping of hazard areas. Again, the following points are central to this concept:

(1) *Hazards must be evaluated based on an understanding of coastal physical processes from a geologic point of view.* The barrier island, its dunes, beach, and offshore are all part of one large geobiological system impacted by several different types of processes (e.g., wind, waves, currents, storm surge).

(2) *Recognition of hazard areas is imperative.* By recognizing hazard areas, development can be directed away from inlet hazards, potential overwash zones, low-elevation areas, back-side flood zones, and so forth.

(3) *Approaches to property damage mitigation must be taken in recognition of the fact that sea level is rising.* The present interglacial period is resulting in a worldwide shoreline migration as sea level rises over a sloping land surface. The sea-level rise is likely to continue in the foreseeable future and may accelerate over the next 50 to 100 years due to the greenhouse effect. Storms are a common natural phenomenon and should not be considered "unexpected catastrophes." Storms are the major driving force moving beaches landward. As sea level rises the impact area of storms (hazard zones) moves in a landward direction. Because barrier islands exist only on very flat coastal plains, a small sea-level rise can result

in a large horizontal shoreline retreat. Risk analysis must be updated to take sea-level rise into account.

(4) *Alterations of island environments due to development should be repaired and restored to the natural setting,* especially where the natural protective qualities of the island are reduced. In many cases this will entail little more than restoring relatively small areas to their predevelopment state by rebuilding dunes or replacing maritime vegetation.

(5) *Island sand volume should be augmented or at least maintained.* Emplacing new sand from an off-island source is better than moving sand from place to place on an island. The same is true for native vegetation. Don't lose a grain of sand!

(6) *Potential for property damage must be recognized as both site specific and regional in character.* Each area presents a unique set of circumstances that requires unique solutions, although general principles can be drawn from all of the coastal zone.

(7) *The entire coastal zone (an entire island, for example) must be considered when applying mitigation plans.* Property damage mitigation can no longer be considered appropriate only for the first one or two rows of houses. Likewise, the coastal zone will continue to move landward as the sea level rises. The storm-to-storm crisis approach should be replaced with a search for long-term solutions to this long-term problem.

Hazards, Economics, and Politics

Nature is not the only arena of the coastal zone, and hazardous processes are not the only performers. Property owners, planners, and public officials can mitigate the impact of hazards, but their theater is one of politics and economics, usually governed by their own set of rules. Our survey of coastal communities reveals as great a diversity of attitudes and responses to hazards as the variability in nature. Nevertheless, enough similarity among communities exists to make the following generalizations:

(1) *Development sites are chosen on the basis of market forces, not nature's forces.* Most town sites came into existence without hazard planning. Plantations, port facilities, church camps, hunting clubs, and ultimately the resorts they became evolved haphazardly, with a few exceptions (e.g., Kiawah Island, South Carolina; Sanibel Island, Florida). Such sites on the barrier island shores, along estuaries, on deltas, floodplains, or in seismic zones (e.g., the Boston, Massachusetts, area; the Charleston, South Carolina, area), were often the most hazardous. Barrier island towns, in particular, were platted in traditional grids over fragile, dynamic environments rather than developing with site stability, suitability, and low risk in mind.

(2) *In old developments residents learned from experience: low-risk sites tended to be developed first, leaving high-risk sites/areas to accommodate growth* (e.g., the New Jersey shore; Nags Head, North Carolina; Pawleys Island, South Carolina; Dauphin Island, Alabama). Such development, when threatened, often opted for engineering "solutions" rather than relocation. These stabilization projects are increasingly expensive and often ineffective. For the oldest resorts (e.g., the New Jersey shore; western Long Island, New York) the end results were large walls, little or no beaches, and a kind of coastal urban blight.

(3) *Politicians, and/or the political pressures to which they react, are oriented toward giving priority to economic development/management,* not *protecting the inhabitants.* Development is seen as progress, a way to increase the tax base, and, as in the pioneer days, such progress is still considered our manifest destiny.

(4) *"Protective" regulations to reduce natural hazards are often viewed as threatening to developers as well as some property owners* (e.g., prohibitions on fill and development of wetlands). Developers often resist regulations designed to protect property owners or building occupants.

(5) *Politicians are drawn from the economic community.* They are often owners of undeveloped acreage, developers, suppliers of materials, lawyers, businesspeople and professionals who benefit from growth and development. Even if no conflict of interest is intended, they have a stake in development to protect. So the approval of a new development may be influenced by the property available from a board member or by the vision of lumber sales or restaurant patrons, rather than an evaluation of the development's hazards and risk potential.

(6) *Politicians are the employers, while the day-to-day work is carried out by the employees: the hired town manager, planner, and community development personnel.* These employees by and large do an excellent job for coastal communities. They are knowledgeable, realistic, committed public servants, but they answer to the elected politicians, not the general public.

(7) *When disasters do strike, we depend on firefighters and police as our first line of defense.* The people with the greatest responsibility for public safety sometimes seem to be the least appreciated: we rely on volunteers rather than a salaried firefighting staff; media comments suggest police are overpaid (the high salaries resulting from overtime due to understaffing). The dedication and fearlessness of these brave people during hazardous and extremely stressful times often means the difference between life and death for hundreds or thousands of persons impacted by natural disasters.

(8) *Collective community attitudes are widely variable.* For example, coastal communities that are suburbs or part of larger urban

10.4 A house on Folly Beach, SC, whose owner is proudly displaying a Yard of the Month sign. No better example of misunderstanding of living by the rules of the sea can be found. Here trees were removed and dunes leveled to provide space for a lawn just like one you could find in an upscale suburb of Indianapolis. The community encourages this increasing risk level by awarding those foolish enough to ignore the rules of the sea.

10.5 The cross-dune walkway in front of the Holiday Inn on Jekyll Island, GA. This walkway, like the lawn in figure 10.4, won an award. This award, however, was given for living by the rules of the sea, not for violating them!

complexes have a high number of permanent residents. Their perceived planning needs differ from communities with more transient populations, communities where much of the property is "recreational" or second homes and only a few of the many property owners are permanent voting residents. Newcomers in the latter communities are not experienced with the surprises of shoreline or coastal living. Their attitude is different with respect to planning, and they are often more likely to locate in newer, higher risk developments. When they see a need for mitigation programs, they lack

the political power to influence planning.

(9) *Developers are in business to make money, not to protect the public.* The emphasis is on build-and-sell, not analysis of site-specific risk or islandwide mitigation or future relocation. The construction industry prospers in the post-storm rush to rebuild.

(10) *Banks and other lenders do have a stake in property mortgages.* In some cases, lenders can be a source of information on risk; however, if your credit is good you can get the loan in spite of the risk. You will, however, be required to have federal flood insurance. The

Principles of Damage Mitigation **169**

post–Hurricane Andrew experience suggests that other forms of property insurance are going to be either more difficult to obtain or more expensive for property in the coastal zone.

(11) *Catastrophes often set the stage for bigger catastrophes.* Post-catastrophe "recovery" is a time of shock and haste to put things right again. Rather than a time of careful relocation and risk reduction, houses and multihousing units are rebuilt "bigger and better" in the same high-risk zones. Big catastrophes often bring new and/or stronger regulations: higher insurance rates, upgraded building codes, prohibitions or restrictions on future development, and mitigation against recurrence of the hazard. Expect it.

(12) *The levels of management regulation and politics are as diverse as the communities and hazards* (e.g., town/city, county, state, and federal regulatory agencies, and special regulations for private and public lands such as units of the Nature Conservancy, Heritage Trust, Coastal Barrier Improvement Act [COBIA], Wildlife Preserves, and other restricted lands).

The individual citizen and property owner is the final decision maker. The foregoing chapters provide a starting point to identify and evaluate hazards in terms of potential risk for both property damage and human health and safety. Both individual and collective actions are needed to mitigate property damage. The best actions are those that mimic nature and that address the entire island or coastal zone, not just the shoreline (figs. 10.4 and 10.5). Time is not on the side of the growing investment in property and infrastructure within the coastal zone. Unless systematic mitigation is utilized, losses will escalate in three areas: lives, property, and the coastal environments we came to in order to enjoy.

We do not recommend living on barrier islands and we would not want our loved ones to live there. The hazards are numerous and difficult to avoid by evacuation. Development on them is destroying a crucial and limited ecosystem. However, if you choose to live by the sea, we hope we have convinced you to live by the rules of the sea.

Appendix: Agencies Involved in Coastal Issues

Climatological/Meteorological

Institute for Disaster Research
Texas Tech University
P.O. Box 4089
Lubbock, TX 79409-4089

National Association of Flood and
Stormwater Management Agencies
1225 Eye Street, NW
Washington, DC 20005
(202) 682-3761 ext. 204

National Center for Atmospheric
Research
P.O. Box 3000
Boulder, CO 80307
(303) 497-1215

National Climatic Data Center
National Oceanic and Atmospheric
Administration
Federal Building
Asheville, NC 28801
(704) 259-0344

National Hurricane Center
1320 South Dixie Highway
Room 631
Coral Gables, FL 33146
(305) 666-4612

National Oceanic and Atmospheric
Administration
National Weather Service
Silver Springs, MD 20910
(301) 427-7523

Hurricane Research Center
Atlantic Oceanographic and
Meteorological Laboratory
4301 Rickenbacker Causeway
Miami, FL 33149
(305) 361-4319

Coastal Management

National Oceanic and Atmospheric
Administration
Office of Ocean and Coastal
Resource Management
Coastal Zone Management
Information Center
1825 Connecticut Avenue, NW,
Room 729
Washington, DC 20235
(202) 673-5115

U.S. Army Corps of Engineers
Planning Division
Civil Works Directorate
Office of the Chief Engineer
Washington, DC 20314
(202) 272-0169

Geology/Coastal/Wetlands

Association of State Wetlands
Managers
P.O. Box 2463
Berne, NY 12023
(518) 872-1804

Fish and Wildlife Service
1-800-USA-MAPS

National Sea Grant Depository
Pell Library-Bay Campus
University of Rhode Island
Narragansett, RI 02882

U.S. Geological Survey
421 National Center
Reston, VA 22092
(703) 648-5684

Construction

International Association of
Structural Movers
P.O. Box 268
Weeksport, NY 13166

National Institute of Standards
and Technology
National Engineering Laboratory
Center for Building Technology
Gaithersburg, MD 20899
(301) 975-5904

Emergency Preparedness and Response

American Red Cross
431 18th Street, NW
Washington, DC 20006-5399
(202) 639-3400

Environmental Protection Agency
401 M Street, SW
Washington, DC 20460
(202) 475-8400

Federal Emergency Management
Agency
500 C Street, SW
Washington, DC 20472
(202) 646-2500

Natural Disaster Research

National Research Council
2101 Constitution Avenue
Washington, DC 20418
(202) 334-3312

Natural Hazards Research and
Applications Information Center
Campus Box 482
University of Colorado
Boulder, CO 80309-0482
(303) 492-6818

Also contact state agencies (a call to
your state capital's information line
should properly direct you) for the
following:
State climatologists
Coastal zone management
Emergency management
Geological survey
Natural resources
Water resources

References

Barbour, Michael G., Theodore M. De Jong, and Bruce M. Pavlik. 1985. Marine Beach and Dune Plant Communities. In The *Physiological Ecology of North American Plant Communities*. Ed. Brian F. Chabot and Harold A. Mooney. New York: Chapman and Hall. 296–322.

Barnes, Jay. 1995. *North Carolina's Hurricane History*. Chapel Hill: University of North Carolina Press.

Beatley, Timothy, David J. Brower, and Anna K. A. Schwab. 1994. *Intoduction to Coastal Management*. Washington, DC: Island Press.

Bourdeau, P. F., and H. J. Oosting. 1959. The Maritime Live Oak Forest in North Carolina. *Ecology* 40:148–152.

Brennan, James W. 1991. Meteorological Summary of Hurricane Hugo. In *Impacts of Hurricane Hugo: September 10–22, 1989*. Ed. C. W. Finkl and O. H. Pilkey Jr. *Journal of Coastal Research* special issue 8:1–12.

Bretschneider, C. L. 1966. Wave Generation by Wind, Deep and Shallow Water. In *Estuary and Coastline Hydrodynamics*. Ed. Arthur T. Ippen. New York: McGraw-Hill Book Company, Inc. 133–196.

Broome, S. W., W. W. Woodhouse Jr., and E. D. Seneca. 1982a. *Planting Marsh Grasses for Erosion Control*. Raleigh, NC: University of North Carolina Sea Grant, Bulletin #UNC-SG-81-09.

———. 1982b. *Building and Stabilizing Coastal Dunes with Vegetation*. Raleigh, NC: University of North Carolina Sea Grant, Bulletin #UNC-SG-82-05.

Bryant, E. A. 1991. *Natural Hazards*. Cambridge: Cambridge University Press.

Bush, David M. 1991. Impact of Hurricane Hugo on the Rocky Coast of Puerto Rico. In *Impacts of Hurricane Hugo: September 10–22, 1989*. Ed. C. W. Finkl and O. H. Pilkey Jr. *Journal of Coastal Research* special issue 8:49–67.

———. 1994. Coastal Hazard Mapping and Risk Assessment, National Committee on Property Insurance (now the Insurance Institute for Property Loss Reduction). *Proceedings of the 1993 Annual Forum*. December 9, 1993, San Francisco, CA. 19–26.

Bush, David M., and Orrin H. Pilkey Jr. 1994. Mitigation of Hurricane Property Damage on Barrier Islands: A Geological View. In *Coastal Hazards: Perception, Susceptibility and Mitigation*. Ed. C. W. Finkl Jr. *Journal of Coastal Research*, special issue 12:311–326.

Bush, David M., Richard M. T. Webb, José González Liboy, Lisbeth Hyman, and William J. Neal. 1995. *Living with the Puerto Rico Shoreline*. Durham, NC: Duke University Press.

Clayton, T. D. 1991. Beach Replenishment Activities on U.S. Continental Pacific Coast. *Journal of Coastal Research* 7(4): 1195–1210.

Coch, N. K., and M. P. Wolff. 1991. Effects of Hurricane Hugo Storm Surge in Coastal South Carolina. In *Impacts of Hurricane Hugo: September 10–22, 1989*. Ed. C. W. Finkl and O. H. Pilkey Jr. *Journal of Coastal Research*, special issue 8:201–228.

Culliton, Thomas J., Maureen A. Warren, Timothy R. Goodspeed, Davida G. Remer, Carol M. Blackwell, and John McDonough III. 1990. *Fifty Years of Population Growth Along the Nation's Coasts, 1960–2010*. Rockville, MD: National Oceanic and Atmospheric Administration.

Davis, Robert E., and Robert Dolan. 1993. Nor'easters. *American Scientist* 81:428–439.

Dixon, K., and O. H. Pilkey Jr. 1991. Summary of Beach Replenishment on the U.S. Gulf of Mexico Shoreline. *Journal of Coastal Research* 7(1):249–256.

Dolan, Robert, and Robert E. Davis. 1992. An Intensity Scale for Atlantic Coast Northeast Storms. *Journal of Coastal Research* 8(4): 840–853.

Fisher, J. J. 1962. *Geomorphic Expression of Former Inlets along the Outer Banks of North Carolina*. Master's thesis, University of North Carolina, Chapel Hill.

Gayes, P. T. 1991. Post–Hurricane Hugo Nearshore Side Scan Sonar Survey: Myrtle Beach to Folly Beach, South Carolina. In *Impacts of Hurricane Hugo: September 10–22, 1989*. Ed. C. W. Finkl and O. H. Pilkey Jr. *Journal of Coastal Research*, special issue 8:95–112.

Godfrey, P. J. 1976. Barrier Beaches of the East Coast. *Oceanus* 19(5):27–40.

Godfrey, P. J., and M. M. Godfrey. 1976. *Barrier Island Ecology of Cape Lookout National*

Seashore and Vicinity, North Carolina. National Park Service Scientific Monograph Series no. 9.

Godschalk, D. R., D. J. Brower, and T. Beatley. 1989. *Catastrophic Coastal Storms: Hazard Mitigation and Development Management.* Durham, NC: Duke University Press.

Golden, J. H. 1990. *Meteorological Data from Hurricane Hugo.* Proceedings, 22 Joint UJNR Panel Meetings: Wind and Seismic Effects, May 14–18, 1990.

Gray, W. M. 1992. Predicting Atlantic Seasonal Hurricane Activity 6–11 Months in Advance. *Weather and Forecasting* 7(3):440–455.

Hall, M. J., and O. H. Pilkey Jr. 1991. Effects of Hard Stabilization on Dry Beach Width for New Jersey. *Journal of Coastal Research* 7:771–785.

Hall, M. J., R. S. Young, E. R. Thieler, R. D. Priddy, and O. H. Pilkey Jr. 1990. Shoreline Response to Hurricane Hugo. *Journal of Coastal Research* 6:211–221.

Hayes, Miles O. 1979. Barrier Island Morphology as a Function of Tidal and Wave Regime. In *Barrier Islands, from the Gulf of St. Lawrence to the Gulf of Mexico.* Ed. Stephen P. Leatherman. New York: Academic Press. 1–28.

Hebert, P. J. and G. Taylor. 1988. *The Deadliest, Costliest, and Most Destructive U.S. Hurricanes of the Century.* Coral Gables, FL: National Hurricane Center, Technical Memorandum.

Hobbs, C. H. 1988. *Historical Overview of Federal Beach Nourishment Projects in Florida.* Paper presented at Beach Technology '88, Gainesville, FL.

Howard, James D., Wallace Kaufman, and O. H. Pilkey Jr., eds. 1985. *National Strategy for Beach Preservation.* Proceedings of the Second Skidaway Institute of Oceanography Conference on America's Eroding Shoreline, Savannah, GA.

Kahn, J. K. 1986. Geomorphic Recovery of the Chandeleur Islands, Louisiana, after a Major Hurricane. *Journal of Coastal Research* 2(3):337–344.

Kraus, Nicholas C., and O. H. Pilkey Jr., eds. 1988. The Effects of Seawalls on the Beach. *Journal of Coastal Research,* special issue 4.

Lennon, Gered. 1991. The Nature and Causes of Hurricane-Induced Ebb Scour Channels on a Developed Shoreline. In *Impacts of Hurricane Hugo: September 10–22, 1989.* Ed. C. W. Finkl and O. H. Pilkey Jr. *Journal of Coastal Research,* special issue 8:237–248.

Leonard, Lynn A., Tonya D. Clayton, and Orrin H. Pilkey Jr. 1990. An Analysis of Replenished Beach Design Parameters on U.S. East Coast Barrier Islands. *Journal of Coastal Research* 6:15–36.

Lopazanski, M. J. 1987. *The Effects of Development and Forest Fragmentation on the Maritime Forests of Bogue Banks, NC.* Master's project, Duke University School of Forestry and Environmental Studies, Durham, NC.

Lopazanski, Michael J., Jonathan P. Evans, and Richard E. Shaw. 1988. *An Assessment of Maritime Forest Resources on the North Carolina Coast.* Raleigh, NC: North Carolina Department

of Natural Resources and Community Development, Division of Coastal Management.

Lynch, Lisa L., and Stephen B. Benton. 1985. *Potential Inlet Zones on the North Carolina Coast from Virginia to Cape Hatteras.* Unpublished report prepared for the North Carolina Department of Health, Environment, and Natural Resources, Division of Coastal Management, Raleigh, NC.

McNinch, J. E. 1989. *The Effectiveness of Beach Scraping as a Method of Erosion Control: Topsail Beach, North Carolina.* Master's thesis, Institute of Marine Science, University of North Carolina, Chapel Hill.

Mercado, Aurelio. 1995. On the Use of NOAA's Storm Surge Model, SLOSH, in Managing Coastal Hazards—The Experience in Puerto Rico. *Natural Hazards* 10(3):235–241.

Morton, R. A., J. G. Paine, D. J. Adilman, and J. A. Digiulio. 1985. *Beach and Vegetation-Line Changes at Galveston Island, Texas: Erosion, Deposition, and Recovery from Hurricane Alicia.* Austin: University of Texas, Bureau of Economic Geology, Geological Circular no. 85-5.

NCDCM. (North Carolina Division of Coastal Management). 1992. *Long-Term Erosion Rates Updated Through 1992.* Raleigh, NC: North Carolina Department of Environment, Health, and Natural Resources.

NCDEM. (North Carolina Division of Emergency Management). 1987. *Eastern North Carolina Hurricane Evacuation Study.* Appendix A,

Inundation Maps. Raleigh, NC: North Carolina Division of Emergency Management.

Neumann, C. J., G. W. Cry, E. L. Caso, and B. R. Jarvinen. 1989. *Tropical Cyclones of the North Atlantic Ocean, 1871–1986*. National Climatic Data Center, Asheville, NC, in cooperation with the National Hurricane Center. National Oceanic and Atmospheric Administration, Historical Climatology Series 6-2 (updated yearly).

NOAA/National Ocean Service. 1990. *Data Report: Effects of Water Levels and Storm Surge Recorded at NOAA/NOS Water Level Stations*. Silver Spring, MD: U.S. Department of Commerce, National Oceanic and Atmospheric Administration.

Nordstrom, K. F. 1987. Shoreline Changes on Developed Coastal Barriers. In *Cities on the Beach: Management Issues of Developed Coastal Barriers*. Ed. R. H. Platt, S. G. Pelczarski, and B. K. R. Burbank. Chicago: University of Chicago, Department of Geography, Research Paper no. 224.

Nordstrom, K. F., N. P. Psuty, and R. W. G. Carter. 1990. *Coastal Dunes: Form and Process*. New York: Wiley and Sons.

NRC (National Research Council). 1994. *Facing the Challenge: The U.S. National Report to the International Decade of Natural Hazard Reduction World Conference of Natural Disaster Reduction*. Yokohama, Japan, May 23–27, 1994. Washington, DC: National Academy Press.

NWS (National Weather Service). 1993. *Hurricane Andrew: South Florida and Louisiana, August 23–26, 1992*. Silver Spring, MD: U.S. Department of Commerce, National Oceanic and Atmospheric Administration, Natural Disaster Survey Report.

Parchure, 1982. *St. Marys Entrance*. Glossary of Inlets Report #11, University of Florida, Gainesville.

Penland, S., D. Nummedal, and W. E. Schramm. 1980. Hurricane Impact at Dauphin Island, Alabama. In *Coastal Zone '80, Proceedings of the Second Symposium on Coastal and Ocean Management*. Vol 2. Ed. B. L. Edge. New York: American Society of Civil Engineers. 1425–1447.

Pilkey, Orrin H., Jr. 1988. A "Thumbnail Method" for Beach Communities: Estimation of Long-term Beach Replenishment Requirements. *Shore and Beach* 56:25–31.

Pilkey, Orrin H. Jr. 1991. Coastal Erosion. *Episodes* 14(1):46–51.

Pilkey, O. H. Jr., and Tonya D. Clayton. 1989. Summary of Beach Replenishment Experience on U.S. East Coast Barrier Islands. *Journal of Coastal Research* 5(2):147–159.

Pilkey, O. H. Jr., and W. J. Neal. 1980. Barrier Island Hazard Mapping. *Oceanus* 23(4):38–46.

———. 1988. Coastal Geologic Hazards. In *The Geology of North America*. Vols 1–2. *The Atlantic Continental Margin*. Ed. R. E. Sheridan and J. A. Grow. Geological Society of America.

Pilkey, O. H. Jr., W. J. Neal, O. H. Pilkey Sr., and S. R. Riggs. 1982. *From Currituck to Calabash, Living with North Carolina's Barrier Islands*. Durham, NC: Duke University Press.

Pilkey, Orrin H. Jr., Dinesh C. Sharma, Harold R. Wanless, Larry J. Doyle, Orrin H. Pilkey Sr., William J. Neal, and Barbara L. Gruver. 1984. *Living With the East Florida Shore*. Durham, NC: Duke University Press.

Pilkey, Orrin H., Jr., and Howard L. Wright III.

1988. Seawalls Versus Beaches. In *The Effects of Seawalls on the Beach*. Ed. N. C. Krause and O. H. Pilkey Jr. *Journal of Coastal Research*, special issue 4:41–64.

Priddy, R. D. 1991. *Effects of Storm-Surge Ebb on South Carolina Barrier Island Coastal Development*. Master's thesis. Department of Geology, Duke University, Durham, NC.

Psuty, Norbet P. 1988. Sediment Budget and Dune/Beach Interaction. *Journal of Coastal Research*, special issue 3:1–5.

SCCC (South Carolina Coastal Council; now the South Carolina Department of Health and Environmental Control, Office of Ocean and Coastal Resource Management, OCRM). Undated. *How to Build a Dune*. Columbia: South Carolina Coastal Council.

Schuck-Kolben, R. Erik. 1990. *Storm-Tide Elevations Produced by Hurricane Hugo along the South Carolina Coast, September 21–22, 1989*. U.S. Geological Survey Open-File Report OF 90-0386, prepared in cooperation with the Federal Emergency Management Agency.

SCMEP (South Carolina Marine Extension Program). March 1990. *Restoring and Maintaining South Carolina's Sand Dunes*. Charleston, Conway, and Georgetown Cooperative Extension Service, Clemson University. Leaflet 40.

Seltz, J. 1976. *The Dune Book: How to Plant Grasses for Dune Stabilization*. Raleigh, NC: University of North Carolina Sea Grant.

Simpson, R. H. 1974. The Hurricane Disaster Potential Scale, *Weatherwise* 27: 169, 186.

Simpson, R. H., and H. Riehl. 1981. *The Hurricane and Its Impact*. Baton Rouge: Louisiana State University Press.

Stanczuk, D. T. 1975. *Effects of Development on Barrier Island Evolution, Bogue Banks, North Carolina*. Master's thesis. Duke University, Department of Geology, Durham, NC.

Stevens, R. A. 1960. *Amelia Island, Florida, Beach Erosion Control Study*. Washington, DC: U.S. Government Printing Office.

Stringer-Robinson, G. 1989. *Time and Tide on Folly Beach, South Carolina (a History)*. Folly Beach, SC: Self-published.

Sylvia, David M. 1989. Nursery Inoculation of Sea Oats with Vesicular-Arbuscular Mycorrhizal Fungi and Outplanting Performance on Florida Beaches. *Journal of Coastal Research* 5(4): 747–754.

Texas General Land Office, 1991. *Dune Protection and Improvement Manual for the Texas Gulf Coast*. Austin: Texas General Land Office, Resource Management and Development.

Thieler, E. R., and D. M. Bush. 1991. Gilbert and Hugo: Hurricanes with Powerful Messages for Coastal Development. *Journal of Geological Education* 39:291–299.

Thieler, E. R., D. M. Bush, and O. H. Pilkey Jr. 1989. Shoreline Response to Hurricane Gilbert: Lessons for Coastal Management. *Coastal Zone '89, Proceedings of the Sixth Symposium on Coastal and Ocean Management*. New York: American Society of Civil Engineers. 765–775.

Thieler, E. R., and R. S. Young. 1991. Quantitative Evaluation of Coastal Geomorphic Changes in South Carolina after Hurricane Hugo. In *Impacts of Hurricane Hugo: September 10–22, 1989*. Ed. C. W. Finkl and O. H. Pilkey Jr. *Journal of Coastal Research*, special issue 8:187–200.

UASCOE (U.S. Army Corps of Engineers). 1984. *Shore Protection Manual*. 3 vols. Stock no. 008-022-00218-9. Washington, DC: U.S. Government Printing Office.

U.S. Dept. of Commerce. 1990. *Natural Disaster Survey Report: Hurricane Hugo, September 10–22, 1989*. Silver Spring, MD: National Oceanic and Atmospheric Administration, National Weather Service.

Wakulla County Planning Board. 1994. *Wakulla County Comprehensive Plan*. Crawfordsville, FL: Wakulla County Planning Board.

Watson, Ben. 1993. New Respect for Nor'easters. *Weatherwise* 46(6):18–23.

Wells, John T., and J. McNinch. 1991. Beach Scraping in North Carolina with Reference to Its Effectiveness During Hurricane Hugo. In *Impacts of Hurricane Hugo: September 10–22, 1989*. Ed. C. W. Finkl and O. H. Pilkey Jr. *Journal of Coastal Research*, special issue 8:249–262.

Williams, Jack. 1992. *The Weatherbook*. New York: Vintage Books.

Williams, John M. 1993. *An Examination of the Risks of Coastal Development and Some Possible Mitigation Methods: A Case Study of Nags Head, North Carolina*. Master's project.

Duke University, School of the Environment, Durham, NC.

Woodhouse, W. W., Jr. 1978. *Dune Building and Stabilization with Vegetation*. Stock no. 008-022-00124-7. Washington, DC: U.S. Government Printing Office,

Wright, L. D., F. J. Sway, and J. M. Coleman. 1970. *Effects of Hurricane Camille on the Landscape of the Breton-Chandeleur Island Chain and the Eastern Portion of the Lower Mississippi Delta*. Baton Rogue: Louisiana State University, Coastal Studies Institute Technical Report 76.

Zarillo, G. A., L. G. Ward, and M. O. Hayes. 1985. *An Illustrated History of Tidal Inlet Changes in South Carolina*. Charleston, SC: South Carolina Sea Grant Consortium.

Index

David M. Bush is Assistant Professor of Geology at
West Georgia College. Orrin H. Pilkey Jr. is Professor
of Geology at Duke University. William J. Neal is
Professor of Geology at Grand Valley State University.

Bush, David M.
Living by the rules of the sea / David M. Bush,
Orrin H. Pilkey Jr., William J. Neal.
p. cm. —(Living with the shore)
Includes bibliographical references and index.
ISBN 0-8223-1801-6 (cl: alk. paper).
ISBN 0-8223-1796-6 (pa: alk. paper)
1. Coastal zone management. 2. Hurricane protection.
3. Barrier islands. 4. Shore protection. I. Pilkey, Orrin
H. Jr., 1934– . II. Neal, William J. III. Title. IV. Series.
HT391.B83 1996
333.91'7—dc20   95-50855 CIP